Progress in Theoretical Computer Science

Guo-Qiang Zhang

Logic of Domains

Birkhäuser 1991
Boston • Basel • Berlin

Guo-Qiang Zhang
Department of Computer Science
University of Georgia
Athens, GA 30602

Library of Congress Cataloging-in-Publication Data

Zhang, Guo-Qiang, 1960-
Logic of Domains / Guo-Qiang Zhang.
p. cm. -- (Progress in theoretical computer science)
Revision of thesis (Ph. D.)-- University of Cambridge, 1989.
Includes bibliographical references and index.
ISBN 0-8176-3570-X (hard : acid-free) : $49.50. -- ISBN 3-7643-3570-X (hard : acid-free)
1. Programming languages (Electronic computers) --Semantics.
2. Logic, Symbolic and mathematical. I. Title. II. Series.
QA76.7.Z47 1991 91-25464
005. 13--dc20 CIP

Printed on acid-free paper.

ISBN 0-8176-3570-X
ISBN 3-7643-3570-X

Camera-ready text prepared in LaTeX by the author.
Printed and bound by Edwards Brothers, Inc., Ann Arbor, Michigan.
Printed in the U.S.A.

9 8 7 6 5 4 3 2 1

Contents

Acknowledgements

This monograph is a revision of the author's Ph.D. thesis submitted to the University of Cambridge in May, 1989. I am grateful to my thesis supervisor, Glynn Winskel, for his invaluable advice and suggestions, which have had a fundamental influence on the development of the work. Special thanks go to Martin Hyland, not only for his suggestions of deep insight, but also for his generosity and friendliness. I would like to thank Samson Abramsky for reading my thesis with rigor (all possible mistakes remain to be mine, however), and for his enlightening comments which helped improve the presentation.

My stay at Cambridge was made possible by a Research Studentship from Trinity College, for which I am very grateful. Many thanks are due to the Department of Computer Science of Aarhus University for its hospitality and for providing me the opportunity for an enjoyable visit.

This monograph would not have come about without the faithful support of my wife Jiayang. I dedicate it to her.

Preface

This monograph studies the logical aspects of domains as used in denotational semantics of programming languages. Frameworks of domain logics are introduced; these serve as foundations for systematic derivations of proof systems from denotational semantics of programming languages. Any proof system so derived is guaranteed to agree with denotational semantics in the sense that the denotation of any program coincides with the set of assertions true of it. The study focuses on two categories for denotational semantics: SFP domains, and the less standard, but important, category of stable domains.

The intended readership of this monograph includes researchers and graduate students interested in the relation between semantics of programming languages and formal means of reasoning about programs. A basic knowledge of denotational semantics, mathematical logic, general topology, and category theory is helpful for a full understanding of the material.

Part I

SFP Domains

Chapter 1

Introduction

This chapter provides a brief exposition to domain theory, denotational semantics, program logics, and proof systems. It discusses the importance of ideas and results on logic and topology to the understanding of the relation between denotational semantics and program logics. It also describes the motivation for the work presented by this monograph, and how that work fits into a more general program. Finally, it gives a short summary of the results of each chapter.

1.1 Domain Theory

Programming languages are languages with which to perform computation. They range from the more theoretical languages like λ-calculi with various evaluation strategies, Milner's CCS [Mi80] and Hoare's CSP [Ho78], to conventional languages like Pascal and Lisp. The development of high level programming languages has made it necessary to give precise meanings to the syntaxes. The most successful and well-known approach to the semantics of programming languages is due to Scott and Strachey [St64], [ScSt71]. Their idea was that the semantics of a programming language should be formally specified in terms of a small number of basic mathematical constructions on partial orders of information. The mathematical part of their approach is called domain theory, and the methodology which uses domain theory to specify the meaning of a programming language is called denotational semantics.

A denotational semantics of a programming language is usually given by assigning to each piece of program an element in a domain. Due to the possible self-applicative nature of some programs, the domains must sometimes have special properties. A typical requirement is that a domain be isomorphic to its own function space. This is impossible with a naive set construction because exponentiation always increases cardinality on non-trivial sets. Scott's idea was not to work with directly the values and objects of computation, but rather with information about computation. In the framework of complete partial orders and continuous functions, Scott has shown that it is possible to find non-trivial solutions to equations such as

$$D \cong [D \to D],$$

so that denotational semantics can be given to the un-typed λ-calculus, a typical system of self-application. Nowadays domain theory has advanced to the extent that it is possible to give denotational semantics to most programming languages.

A complete partial order (cpo) is a partial order which has a bottom element and least upper bounds of ω-increasing chains. For two elements x and y of a cpo, $x \sqsubseteq y$ means the information content of x is contained in the information content of y. The bottom element $\perp$ has, therefore, empty information content. Elements which have finite information content play an important role in the theory. Intuitively those elements have information which can be realized by a computation in a finite amount of time. We call them finite elements. As far as computation is concerned it is usually enough to work within ω-algebraic cpos, which have a countable number of finite elements as its base. A function between complete partial orders D and E can be seen as a computation which makes use of some input information in D and produces some output information in E. According to Scott's thesis, such a function should be continuous. Continuity requires that the function be monotonic — more information as input yields more information as output — and, if

$$x_0 \sqsubseteq x_1 \cdots \sqsubseteq x_i \sqsubseteq \cdots$$

is a chain in D, the output information for $f(\bigsqcup_{i\in\omega} x_i)$ should be able to be approximated by $f(x_i)$'s as closely as one requires. An important feature of cpos is that the collection of continuous functions on cpos again form a cpo, with the pointwise order; this means that functions themselves are associated with elements of information, which enables the treatment of higher order computations. Unfortunately continuous functions on ω-algebraic cpos need not form an ω-algebraic cpo again. There are, however, subcategories of ω-algebraic cpos with this closure property. The most well-known framework is the category of Scott domains, consisting of as objects those ω-algebraic cpos which are consistently complete. Scott domains have many nice properties. They are closed under sum, product, lifting, and function space (with continuous functions they form a cartesian closed category).

When it comes to specifying the semantics of parallel programming languages, powerdomain constructions [Pl76], [Sm78] are often used. Powerdomains resemble the powerset whose elements represent the 'sets' of different courses a nondeterministic computation can follow. However there are several possible ways to order the elements of a powerdomain; different orderings yield different powerdomains. There are the Hoare powerdomain, the Smyth powerdomain and the Plotkin powerdomain, based on, respectively, three views about what kind of information should be taken into account for a non-deterministic process. There is no problem with the Hoare and the Smyth powerdomain — within the framework of Scott domains they produce a Scott domain from a Scott domain. The Plotkin powerdomain, however, does not produce a Scott domain from a Scott domain. This led Plotkin to the discovery [Pl76] of a more general category of SFP objects (or SFP domains), which are special kinds of ω-algebraic cpos closed under sum, product, lifting, function space, and the three powerdomain constructions just mentioned.

There are many other frameworks for denotational semantics (see Meyer [Me88] for a survey), among which there is the less standard but important category of dI-domains (or stable domains). DI-domains were discovered by Berry [Be78] from the study of the full-abstraction problem for typed λ-calculi. They are special kinds of Scott domains which have a more operational nature. The functions between dI-domains are stable functions under an order which takes into account the manner in which they compute. One of the most important aspects of dI-domains is that in them there is a subcategory of coherent spaces brought into popularity by Girard [Gi89]. Coherent spaces have been recently used to model system F [Gi87a]. Most importantly they have triggered the discovery of linear logic [Gi87b]. Moreover, dI-domains can be represented as stable event structures [Wi86], which can be used to model concurrent processes, and to relate the more abstract domains to the more concrete structure of Petri nets [NiPlWi79], [Wi80]. Most recently, it has been shown that there is a monoidal closed category of stable domains which can serve as a model for intuitionistic linear logic [Zh91] (see also Section 6.3).

1.2 Program Logics

Program logics[1] are formal systems for reasoning about properties of programs. The process of reasoning is usually specified by a set of rules, forming a proof system. One of the well-known program logics is the Floyd-Hoare logic (or Hoare logic) [Ho69]. It uses assertions of the form $\{P\}\,C\,\{Q\}$, meaning that if a program C starts at a state with property P and terminates, it terminates at a state with property Q. There are many other kinds of program logics, such as temporal logic [Pn77] and dynamic logic [Pr79].

Two of the most important theoretical issues related to a proof system are its soundness and completeness. Informally soundness means each assertion derived from the proof system is correct while completeness requires that every correct assertion be derived in this way. While soundness of Hoare logic is easy to establish completeness is impossible because of Gödel's incompleteness theorem for Peano arithmetic. However, Cook [Co78] showed that Hoare logic is complete relative to the truths of arithmetic, in the sense that if one were allowed to consult an oracle about truths of arithmetic one can derive all the correct program assertions. There are other results on the completeness of Hoare logics. For a survey see [Apt81], [Apt84].

Recently, a lot of work has been done to extend the axiomatic method of Hoare to parallel programming languages. One of the approaches is due

[1] The words 'proof system' and 'program logic' are often used as synonyms in the literature.

to Owicki and Gries [OwGr76]. The language they considered was the usual while-programs expanded with statements for parallel composition via shared variables, and critical regions. Their proof system uses assertions similar to Hoare triples. However, the situation become much more complicated. Owicki and Gries need interference-freedom for the soundness and auxiliary variables for the completeness of their proof system. Interference-freedom was a condition attached to the proof rule for parallel composition, which says that processes do not interfere with each other. Recently, Brookes [Br85] presented a novel proof system for the same parallel programming language but avoids the use of auxiliary variables and interference-freedom for the proof system. This is achieved by introducing more structure on assertions. The basic idea is to adopt assertions with a structure similar to that of labeled trees, where nodes are attached with predicates. In this way, one can not only describe the input-output property associated with a command, but also keep track of properties of the intermediate steps in the execution of the command.

There are enormously many other kinds of proof systems for programming languages in the literature. Often there are even several proof systems for the same programming language. However, many of the proof systems are complicated or ad hoc, or even incorrect, not fulfilling what is claimed of them.

Faced with such a situation we may ask whether there is any principle on which our proof systems can be based. We may ask whether there is any general method that can guide us in building proof systems. Fortunately, the line of research described in the next section opens a way that may lead to such a principle and methodology.

1.3 Logic and Topology

Much recent work ([Pl80], [Hy81], [Sc81], [Sc82], [Sm83], [Wi82], [Ro86], [Ro87]) has been concerned with the relation between logic and domains or topology. The significance of this line of enquiry was eventually made clear by Abramsky [Ab87]. He showed that it could be regarded as part of a general program for extracting from a suitable semantics a program logic based on a syntax of types, terms, and predicates. This section gives a review of, from the author's own prospect, the brief history of work on predicate transformers, information systems, and logical approaches to domain theory.

1.3.1 Predicate Transformers

Predicate transformers were introduced by Dijkstra [Di76] when he was dealing with the semantics of a simple non-deterministic language of guarded commands. The idea was to regard each program C as a 'predicate transformer' which transforms each assertion (predicate) Q into the weakest precondition $wp(C, Q)$ among P's such that the Hoare triple $\{P\}C\{Q\}$ is valid. On the other hand, semantics can be given to the programming language by considering each program as a state transformation function abstracted from a transition system.

Plotkin [Pl80] showed that for Dijkstra's language of guarded commands these two approaches coincide. Via Smyth powerdomain Plotkin gave an isomorphism between the cpo of predicate transformers and the cpo of state transformation functions for the language of guarded commands.

Later Smyth [Sm83] investigated this connection from a broader topological stand. Smyth pointed out that the above connection is just a special case of the duality between continuous functions $D \xrightarrow{f} E$ and morphisms

$$\Omega(E) \xrightarrow{f^{-1}} \Omega(D)$$

on the Scott topology. Smyth emphasised the computational significance of topological ideas. A topological space X can be taken as a 'data type', with the open sets as 'predicates' and functions between topological spaces as 'computations' (For a more intuitive, computer science prospect of logic and topology, see [Vi88]).

1.3.2 Information Systems

Information systems (neighborhood systems [Sc81] without points) were introduced by Scott [Sc82] initially with the intention of making domain theory accessible to a wider audience. In this representation the idea of *information* is made explicit — each element is seen as a collection of *information quanta.* It gives a logical approach to domain theory, in which properties of domains can be derived from assumptions about the entailments between propositions.

Intuitively, an information system is a structure describing the logical relations between propositions that can be made about computations. It consists of a set of propositions, a consistency predicate and an entailment relation. An information system determines a family of subsets of propositions called its elements. An element consists of a set of propositions that can be truly made about a possible computation. These elements form a Scott domain under inclusion. On the other hand, given a Scott domain, there is an information system which determines a domain isomorphic to the original one. Information systems form a category with the approximable mappings [Sc82] as morphisms. It is equivalent (in the sense of [Mac71]) to the category of Scott domains. Constructions such as product, sum, and function space have been proposed on information systems [Sc82], [LaWi84], corresponding to those on domains. In addition, recursive information systems can be defined based on fixed-point theory [LaWi84].

Information systems fit Smyth's topological view with propositions (predicates) regarded as tokens for open sets. Let X be a finite consistent set of propositions of an information system. Let

$$O(X) = \{ x \mid X \subseteq x \}$$

where x's are elements of the information system. Then $O(X)$ is an open set in the Scott topology of the domain determined by the information system. On the other hand, to get an information system from a Scott domain D, one can take as propositions the prime open sets, and define the entailment by letting

$$\{ O_1, O_2, \cdots, O_m \} \vdash O$$

if and only if $\bigcap_{1 \leq i \leq m} O_i \subseteq O$. The consistency predicate is equally simple:

$$\{ O_1, O_2, \cdots, O_m \} \in Con$$

if and only if $\bigcap_i O_{1 \leq i \leq m} \neq \emptyset$. Such a smooth transformation is guaranteed by properties of algebraic cpos, which include, among other things, that Scott topology on an algebraic cpo is both T_0 and sober (for more detail see [Sm83]).

1.3.3 Domains and Logics

The work on logic of domains has been based, to various extent, on different instances of Stone dualities. Two kinds of categories are involved: a category of complete Heyting algebras, and a category of topological spaces.

Complete Heyting algebras are complete lattices satisfying the infinite distributive law

$$a \wedge \bigvee S = \bigvee \{ a \wedge s \mid s \in S \}.$$

The category of *frames* has objects complete Heyting algebras, and morphisms functions which preserve finite meets and arbitrary joins. As a special kind of frames one has the set of open sets $\Omega(D)$ of a topological space D ordered by inclusion; in this case the frame morphisms are precisely those functions

$$f^{-1} : \Omega(E) \rightarrow \Omega(D)$$

for which $f : D \rightarrow E$ is continuous. The category of *locales* is the opposite of the category of frames. *Stone dualities* are contravariant equivalences between certain categories of topological spaces and the corresponding categories of locales (Johnstone's book [Jo82] is recommended here). Examples of this kinf of dualities have already appeared in the literature. Information systems, for example, can be seen as a description of locales where the relevant topological spaces consist of the Scott open sets. The *duality* between the category of information systems and the category of Scott topologies of domains can be viewed as just the *equivalence* between the category of information systems and the category of domains. There are other results of a similar nature, which we discuss now.

The program logics derived from domains should allow reasoning at higher types. Function spaces in the category of locales has been treated in great detail in [Hy81], where Hyland showed that a locale is exponentiable if and only if it is locally compact. A construction of the function space of locales has been given explicitly.

To treat non-determinism, Winskel [Wi82] observed a simple connection between powerdomains and modal assertions. The modalities are $\Box$, for 'inevitably', and $\Diamond$, for 'possibly', to make modal assertions about non-deterministic computations. Winskel showed that the Smyth powerdomain is built up from assertions about the 'inevitable' behavior of a process, the

Hoare powerdomain is built up from assertions about the 'possible' behavior of a process, while the Plotkin powerdomain is built up from both kinds of assertions taken together.

Powerdomains are closely related to the classical Vietoris construction on topological spaces. Vietoris showed that for a compact Hausdorff space X, there is a compact Hausdorff topology on the set $\mathbf{K}(X)$ of non-empty, closed subsets of X which coincides with that induced by the Hausdorff metric, provided the topology on X is determined by a metric. However, the relevance of Vietoris construction to computer science was neglected until [Sm82], where Smyth pointed out the relationship between powerdomains and Vietoris topology. Given a multifunction $\Gamma : X \to Y$, there are three ways to define continuity: the upper semicontinuity, the lower semicontinuity, and both taken together[2]. Correspondingly three topologies were introduced: the upper topology, the lower topology, and the Vietoris topology. The three notions of continuity for a multifunction $\Gamma : X \to Y$ can then be characterized topologically as continuous functions from X to $\mathcal{P}Y$, where $\mathcal{P}Y$ is a suitable topological space constructed out of Y. Smyth showed that by removing the empty set, the specialization orders determined by the upper power space, lower power space, and convex power space are isomorphic to the Smyth powerdomain, the Hoare powerdomain, and the Plotkin powerdomain, respectively.

A more formal approach which takes powerdomains as constructions on locales was given by Robinson [Ro86]. Robinson showed that a Vietoris locale gives a Scott topology on the Plotkin powerdomain. The presentation of a domain by means of an algebraic description of its lattice of open sets supplies a way to 'transform' domains into sets of proof rules, such as

$$\Diamond\phi \wedge \Box\psi \leq \Diamond(\phi \wedge \psi)$$

and

$$\Box(\phi \vee \psi) \leq \Diamond\phi \vee \Box\psi$$

for the Plotkin powerdomain construction. Such rules have been known to some people [Jo82], and were implicit in Winskel's work [Wi82], but was

[2] A multifunction $\Gamma : X \to Y$ is called upper semicontinuous if $\Gamma^+ : \Omega(Y) \to \Omega(X)$ is a function, where $\Gamma^+(P) = \{ a \mid \Gamma a \subseteq P \}$; and lower semicontinuous if $\Gamma^- : \Omega(Y) \to \Omega(X)$ is a function, where $\Gamma^-(P) = \{ a \mid \Gamma a \cap P \neq \emptyset \}$.

brought to the fore by Robinson [Ro86]. It was also made clear that the finite elements of a Plotkin powerdomain are captured by assertions of the form

$$(\Box \bigvee_{i \in I} \phi_i) \wedge \bigwedge_{i \in I} \Diamond \phi_i.$$

A notable advance in this direction has been made by Abramsky [Ab87] when he showed how program logics could be extracted from domain theory for the case of Scott domains. His work provided successful treatment of morphisms, an important step which showed definite promise of frameworks of this kind since they can, in a sense, generalize and express dynamic logic and Hoare logic. Parallel to the work of Abramsky, Robinson [Ro87] has provided a framework for obtaining axiomatic semantics from denotational semantics of programming languages.

1.4 Logic of Domains

As indicated in the previous sections, most of the work done so far are on the logical aspects of Scott domains. When it comes to stable domains it is important to note the differences between the theory for Scott domains and the theory for stable domains. In both cases we are interested in certain families of subsets. However, while in the case of Scott domains these embed nicely as the basis of a topology, in the case of stable domains, because of their failure to be closed under non-disjoint union they do not, with important consequences for the interpretation of the logic.

This monograph presents the author's work on the logical aspects of SFP domains and stable domains. The intended purposes are to convey to the reader the state of knowledge on logic of domains and to point to areas where more research is to be done.

1.4.1 Motivation

There are three approaches to the semantics of programming languages: operational semantics (or structural operational semantics, natural semantics), denotational semantics, and axiomatic semantics. Operational semantics specifies the behavior of a program by the effect of the stepwise execution of its atomic actions. Closer to what is actually happening in reality, this is the place where many of the computational intuitions and ideas first take their formal shape. Denotational semantics maps programs to computable functions in some mathematical space, or, to some points in a higher order domain. It frees one from the sometimes unwanted detail, so that attention can be focused on issues at a higher level. Axiomatic semantics is most directly related to the correctness of programs. Here the behavior of a program is expressed in an assertion language, and the verification of the correctness is to check that a program meets its specification. Proof systems are usually used for axiomatic semantics. In a generalized sense, however, this may not be required. Axiomatic semantics should be considered as given as long as an assertion language and the definition of when a program satisfies an assertion are defined.

Note that success in the verification of 'correctness' does not guarantee absolute correctness, because the specification may contain human error. However, one should avoid specifying the property of a big program by a big assertion by brute force. Instead, good proof systems should make it possible to break a big program along with its specification into smaller and smaller pieces until the correctness of the smaller programs becomes trivial. This is the so called compositional approach, which would greatly improve the reliability of programs than the testing method.

To be able to take advantage of the different semantic approaches to programming languages, it is important to understand the relationships among them. It would be ideal to be able to ensure that different approaches end up formally equivalent in some sense when needed. The connection between operational semantics and denotational semantics has been well studied, as in *e.g.* full abstraction [Mi77], [BCL85], [St88]. However, not a great deal of attention has been paid to the relationship between denotational semantics and axiomatic semantics. There is the following important issue: for a programming language with a denotational semantics, how can we, if possible, associate it with a proof system so that axiomatic semantics *agrees* with denotational semantics in the sense that for programs C, C', $\{ C \} \subseteq \{ C' \}$ if and only if $[\![C]\!] \sqsubseteq [\![C']\!]$, where $\sqsubseteq$ is the order inherited from the domain, and $\{ C \}$ is the collection of assertions true of C?

The ideas and results about domains and logic mentioned in the previous sections are especially relevant because on one hand, locales have more of a logic flavor and constructions on them provide proof rules; on the other hand, topological spaces can give rise to domains in which programs are given denotational semantics. Stone dualities allow smooth transformation from one framework to another. Accordingly, they are considered as the appropriate bridges with which to connect denotational semantics and program logics. In particular, they ensure that axiomatic semantics agrees with denotational semantics.

As indicated in Section 1.2, however, many of the existing program logics have not been developed to such a maturity as to facilitate the derivation of denotational semantics from them using some form of Stone duality (Hennessy-Milner logic seems to be an exception [Ab87a], however). In contrast, domain theory is well-developed and sophisticated. This leaves us with only one side to explore: to build up logical frameworks from domains and use them for the derivation of program logics.

The work involved can be classified crudely into two kinds. The first kind is to develop logical frameworks for various categories of domains for denotational semantics. What is given here is certain category of complete partial orders. Although sometimes there is a natural candidate for the topology characterizing the morphisms and the complete partial orders (the Scott topology on Scott domains, for example), in other cases such a topology has to be developed (such as in the case of stable domains). What kind of topology should the domains come equipped with may depend on the kind of program logic we want to develop. The appropriate topology selected serves as a basis for the logical framework, with assertions of the logic interpreted as the open sets. It is possible that the logical forms suitable for our purpose have already been studied for some other reason in detail, and in this case we can save a lot of work.

As realized by Abramsky [Ab87], the complete primes in the lattice of Scott open sets play a key role in the logic. There is a meta-predicate that identifies the syntax of assertions whose interpretations are complete primes. The meta-predicate is used as side conditions for some of the proof rules, without which completeness cannot be achieved. This is where the notion of an information system becomes important, because the propositions of information systems correspond to the complete primes of the lattice of Scott open sets. Therefore, information systems are the back bones of the logic.

The second kind of work is about applications of domain logic. Given a programming language, a domain in some category is chosen for denotational semantics. The domain is usually recursively defined, being the solution to some domain equations. The selection of the category and the domain depends on the programming language and the aspects of the computation to be captured. It is not the concern of the monograph how such a selection should be made. However, once a domain is selected the logical framework (the first kind of work) can generate automatically a logic for axiomatic semantics of the programming language.

1.4.2 Summary of Results

Most of the work presented in the monograph is of the first kind. The first part (Chapter 1 to Chapter 5) is concerned with the category of SFP domains, and the second part (Chapter 6 through Chapter 10) with the category of stable domains. A brief summary of the results of each chapter is given below.

Chapter 2 offers some background knowledge needed for the understanding of the work in later chapters. Some of the most often used terminologies in domain theory are given formally, and some known results about powerdomains and SFP objects are stated, but proofs are omitted from time to time. A new characterization of the finite elements in the function space of SFP domains is introduced at the end.

Chapter 3 introduces an information-system representation for SFP domains. A Gentzen style entailment $X \vdash Y$, instead of $X \vdash a$ is adopted. A category of special kinds of such systems is shown to be equivalent to the category of SFP domains. Constructions like the Plotkin powerdomain and the function space are given, as well as a cpo of such systems to give meanings to recursively defined domains. This is a step from SFP domains to simple logical forms, making it smoother to arrive at the logic described in Chapter 4.

Chapter 4 gives a logic of SFP domains. A meta-language for denotational semantics is introduced, with type constructions like sum, product, function space, the three powerdomains, and recursively defined types. For each closed type there is a language of open set assertions. Proof systems are given, and soundness and completeness results are obtained. As an application of the logic, the style of Brookes' assertions is shown to be determined by the logic of Plotkin's domain of resumptions. Finally, a proof system for reasoning about negative information is provided.

Chapter 5 studies a μ-calculus of domain logic to extend the expressive power. A least fixed-point operator is added to the assertion language as the μ-construction, to represent the least fixed-point. The μ-calculus includes proof rules for fixed-point induction. It is shown to be sound. Many theorems are derived from the proof system, which are used in the proof of completeness for a special case — the μ-calculus of integers. As the proof of completeness is achieved by normal form theorems, the expressive power of the integer μ-calculus is immediately clear. A characterization of the definable subsets of $\mathcal{N}_\perp$ is given.

Chapter 6 presents categories of coherent spaces, stable domains, and event structures, to provide background knowledge for the work of the second part.

Chapter 7 introduces prime information systems. Prime information systems are special kind of information systems which represent stable domains. Constructions like lifting, sum, product and function space are proposed. A cpo of prime information systems is given, making it possible to define recursive prime information systems by fixed-point theory. This chapter plays a similar role to Chapter 3. It provides a characterization of complete primes in the lattice of stable neighborhoods introduced in Chapter 8.

Chapter 8 investigates stable neighborhoods. Unlike in the first part where Scott topology is suitable, the 'topology' on stable domains has to be developed. Stable neighborhoods characterize stable functions in a similar sense to that in which Scott-open sets characterize continuous functions. However they do not form a topology in the usual sense — they are not closed under non-disjoint unions. Constructions on stable neighborhoods are given with respect to constructions such as sum, product, tensor product, and stable function space. In the category of coherent spaces we introduce further constructions on stable neighborhoods such as linear function space and exponential. In all these cases the constructions preserve compactness. These constructions suggest proof rules to be used in Chapter 9. Investigation is also made on stable neighborhoods of event structures. Events as well as relations among them are shown to determine stable neighborhoods. The partially synchronous product, however, does not preserve compactness of stable neighborhoods.

Chapter 9 studies the logic of stable domains. Two logical frameworks are introduced, one for $\mathbf{COH}_l$, coherent spaces with linear, stable functions; the other for **DI**, dI-domains with stable functions. The logic of $\mathbf{COH}_l$ can be used to derive a logic for $\mathbf{COH}_s$, coherent spaces with stable functions, via the exponential construction. The assertion language for the logic use a 'disjoint or', corresponding to the disjunctive nature of stable neighborhoods. Because of the disjunctive property, assertions are formulated by syntactic rules and proof rules together. Proof systems are introduced with novel proof rules dealing with different type constructions. Soundness, completeness, and expressiveness results are given. An alternative approach to the logic of stable domains is discussed in the last section.

Chapter 10 summarizes the work and suggests areas where more research is to be done.

Chapter 2

Prerequisites

This chapter gives a concise introduction to domain theory. It sets up notations and reviews results which will be used repeatedly in the sequel. Readers new to the field may find [St77], [Pl82], [Sc86], and [Wi90] helpful.

2.1 Cpos and Domains

A *partial order* is a set D with a binary relation $\sqsubseteq \subseteq D \times D$ which is

- reflexive: $x \sqsubseteq x$ for every x in D,
- transitive: $x \sqsubseteq y \,\&\, y \sqsubseteq z \Rightarrow x \sqsubseteq z$, and
- antisymmetric: $x \sqsubseteq y \,\&\, y \sqsubseteq x \Rightarrow x = y$.

Without the third axiom, $(D, \sqsubseteq)$ is called a *preorder*. When $x \sqsubseteq y$ in a partial order, we say x is below y, or y dominates x.

There is a standard notion of *least upper bounds* for a partial order $(D, \sqsubseteq)$. Let X be a subset of D. Say y is an *upper bound* of X, if y dominates every element in X. The least upper bound (also called supremum, or join) of X is the smallest upper bound of X. Of course if the least upper bound exists, it must be unique. If X consists of two elements a and b, the least upper bound of X is written as $a \sqcup b$. In general, the least upper bound is written as $\bigsqcup X$. But when $X = \{x_0, x_1, \cdots\}$, we sometimes write $\bigsqcup_{i \in \omega} x_i$ for the least upper bound. Here ω is the set $\{0, 1, \cdots\}$.

There is a dual notion called the *greatest lower bound* (also named infimum, or meet). For a subset X of D, y is called an *lower bound* of X if y

is below every element in X. The greatest lower bound of X, as the name suggests, is a lower bound of X which is bigger than any other lower bound of X. It is written as $\bigsqcap X$. We write $a \sqcap b$ for the greatest lower bound of a and b. When $X = \{ x_i \mid i \in \omega \}$, we sometimes write $\bigsqcap_{i \in \omega} x_i$ instead of $\bigsqcap X$.

A subset X of a partial order $(D, \sqsubseteq)$ is called *compatible*, written $X\uparrow$, if there exists an upper bound for X. Otherwise it is called *incompatible*. When $\{ a, b \}$ is compatible we write $a \uparrow b$.

A *directed* set of a partial order $(D, \sqsubseteq)$ is a non-null subset $S \subseteq D$ such that every pair of elements in S has an upper bound in S. A *complete partial order* (cpo) is a partial order $(D, \sqsubseteq)$ which has a least element $\bot$ and all least upper bounds of directed subsets of D. An *isolated (or finite)* element of a cpo $(D, \sqsubseteq)$ is an element $x \in D$ such that for any directed subset $S \subseteq D$, if $x \sqsubseteq \bigsqcup S$ then there is an $s \in S$ such that $x \sqsubseteq s$. We write D^0 for the set of finite elements of D. A cpo $(D, \sqsubseteq)$ is *algebraic* if for all $x \in D$, $\{ e \sqsubseteq x \mid e \in D^0 \}$ is directed and $x = \bigsqcup\{e \sqsubseteq x \mid e \in D^0\}$. *Domains* are algebraic cpos. However we should not take this name too seriously; there have been different definitions of domains in the literature. When $(D, \sqsubseteq)$ is algebraic and D^0 countable, D is called *ω-algebraic.* A cpo is a *Scott domain* if it is ω-algebraic and *consistently complete, i.e.*, every compatible subset has a least upper bound.

Another definition of domains often referred to in the literature is based on ω-increasing chains rather than directed sets. An ω-increasing chain in a partial order $(D, \sqsubseteq)$ is a countable subset $\{ x_1, x_2, \cdots x_n, \cdots \}$ of D such that

$$x_1 \sqsubseteq x_2 \sqsubseteq \cdots \sqsubseteq x_n \cdots.$$

A complete partial order is a partial order $(D, \sqsubseteq)$ which has a least element $\bot$ and all least upper bounds of ω-increasing chains. A finite element of a cpo $(D, \sqsubseteq)$ is an element $x \in D$ such that for any ω-increasing chain $\{ x_i \mid i \in \omega \} \subseteq D$, $x \sqsubseteq \bigsqcup_{i \in \omega} x_i$ implies the existence of some x_n such that $x \sqsubseteq x_n$. A cpo $(D, \sqsubseteq)$ is *algebraic* if for all $x \in D$ there is an ω-increasing chain $\{ x_i \mid i \in \omega.\ x_i \sqsubseteq x \}$ of isolated elements such that $x = \bigsqcup_{i \in \omega} x_i$. When $(D, \sqsubseteq)$ is algebraic and the set of finite elements is countable, D is called *ω-algebraic.* These two definitions, however, coincide for ω-algebraic cpos.

Theorem 2.1 *(Plotkin [Pl78]) A partial order $(D, \sqsubseteq)$ is ω-algebraic in terms of directed subsets if and only if it is so in terms of ω-increasing chains.*

For ω-algebraic cpos, then, directed subsets and ω-increasing chains can be used interchangeably.

Lemma 2.1 *If $X \subseteq D^0$ is such that $\forall x \in D.\ \{ y \in X \mid y \sqsubseteq x \}$ is directed and $x = \bigsqcup \{ y \in X \mid y \sqsubseteq x \}$, then $X = D^0$.*

This lemma says in a cpo if we have some set of finite elements which can be used to obtain all other elements as the least upper bounds in certain way, that set actually contains all the finite elements. Although easy to prove, it will be very useful later when we deal with finite elements under various contexts.

The following are a few examples of domains. Domains are usually drawn as Hasse diagrams with smaller elements below bigger ones, and immediately comparable pairs of elements having an edge between them.

Example 2.1 *The one-point domain $\perp$.*

Example 2.2 *The two-point domain $\mathcal{O}$, also called the Sierpinski space.*

Example 2.3 *The truth value cpo $\mathcal{T}$.*

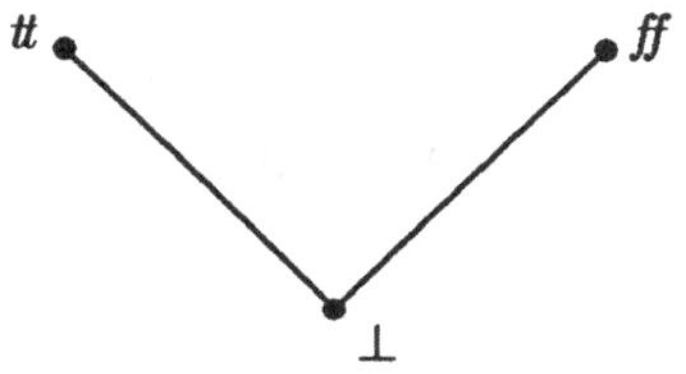

Example 2.4 $\mathcal{N}_\perp$, *the domain of natural numbers.*

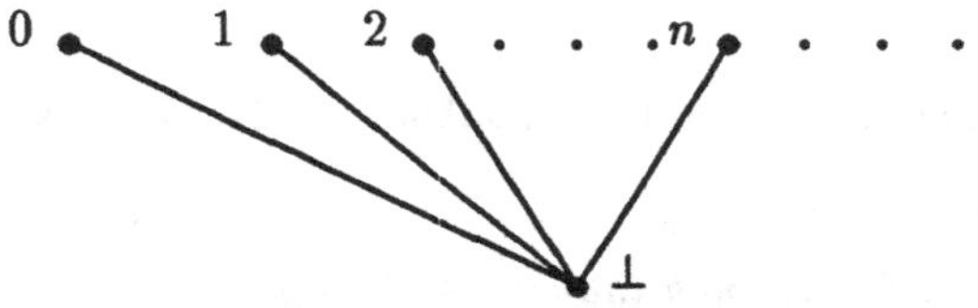

Similarly, by joining a bottom element to any set S, we get a cpo $S_\perp$. Such cpos are called ***flat domains*** for obvious reason.

There is a useful notion stronger than that of finite elements. Call an element p of a cpo D a *complete prime* if whenever $X\uparrow$ and $p \sqsubseteq \bigsqcup X$ for some subset X of D, there is some x in X such that $p \sqsubseteq x$. Note although a bottom is always a finite element, it is never a complete prime. This is because $\bigsqcup \emptyset = \perp$.

2.2 Constructions on Cpos

The standard constructions on domains are function space, product, sum, and lifting. They are most often used in denotational semantics of programming languages.

2.2.1 Function space

Let D, E be cpos. A function $f : D \to E$ is *continuous* if it is monotonic, *i.e.*

$$x \sqsubseteq y \Longrightarrow f(x) \sqsubseteq f(y),$$

and it preserves the least upper bounds of directed sets, *i.e.*

$$f(\bigsqcup X) = \bigsqcup f(X)$$

for any directed subset X of D, where $f(X) = \{f(x) \mid x \in X\}$, to be precise. Actually monotonicity is implied by the preservation of least upper bounds of directed sets. To emphasize monotonicity it is put as an axiom explicitly.

There is a definition of continuous functions in terms of ω-increasing chains. A function is called chain-continuous if it is monotonic and for any chain

$$x_0 \sqsubseteq x_1 \sqsubseteq x_2 \cdots,$$

$$f(\bigsqcup_{i\in\omega} x_i) = \bigsqcup_{i\in\omega} f(x_i).$$

For ω-algebraic cpos, however, these two definitions coincide. It is easy to see that directed-continuity implies chain-continuity. Conversely, suppose a function $f : D \to E$ is chain-continuous. For any directed subset X of D,

$$\bigsqcup X = \bigsqcup\{ d \mid d \in D^0 \ \& \ d \sqsubseteq \bigsqcup X \},$$

by the algebraic property of D. However,

$$\{ d \mid d \in D^0 \ \& \ d \sqsubseteq \bigsqcup X \}$$

is countable because D is ω-algebraic. It is then possible to construct an ω-increasing chain, in X, with limit $\bigsqcup X$. Now, by chain-continuity, the least upper bound of X is preserved by f. In fact we do not need any restriction on E. Therefore,

Theorem 2.2 *Let D be an ω-algebraic cpo, and E any cpo. A function $f : D \to E$ is continuous if and only if it is chain-continuous.*

When only ω-algebraic cpos are concerned (which is the case in later chapters), it does not matter which notion of continuity is adopted.

Cpos with continuous functions form a category. This means, in particular, that the identity functions are continuous, and the composition of continuous functions is continuous. For cpos D and E, the function space $[D \to E]$ consists of continuous functions $f : D \to E$ ordered pointwise (coordinatewise), *i.e.*,

$$f \sqsubseteq g \text{ if } \forall x \in D.\ f(x) \sqsubseteq g(x).$$

The function space $[D \to E]$ is a cpo with the least element the function $\lambda x \in D.\ \bot_E$, sending everything to the bottom. The least upper bounds of ω-increasing chains of continuous functions are also defined pointwise, *i.e.*,

$$\bigsqcup_{i\in\omega} f_i = \lambda x \in D. \bigsqcup_{i\in\omega} f_i(x).$$

A function $f : D \to E$ is *strict* if $f(\bot_D) = \bot_E$. The strict continuous functions of $[D \to E]$ form a cpo as well, written as $[D \to_\bot E]$.

2.2.2 Product

Let $\{ D_i \mid i \in I \}$ be an indexed family of cpos. Their *product*, $\Pi_{i\in I} D_i$, consists of

$$\{ x : I \to \bigcup_{i \in I} D_i \mid \forall i \in I.\, x_i \in D_i \}$$

where we write x_i for $x(i)$, the i-th component of x. Sometimes we write $\langle x_i \rangle_{i \in I}$ for x. The order is defined by $x \sqsubseteq y$ if $\forall i \in I.\, x_i \sqsubseteq y_i$. When $I = \{1, 2\}$ we write $\Pi_{i\in I} D_i$ as $D_1 \times D_2$. $\Pi_{i\in I} D_i$ is a cpo with the bottom element $\lambda i \in I.\, \bot_{D_i}$ and least upper bounds of directed sets of elements determined coordinatewise. There are nature functions associated with the product, which are called *projections*. The j-th projection $\pi_j : (\Pi_{i\in I} D_i) \to D_j$ takes each element x of the product to its j-th component x_j. Projections are continuous. There are also the *tupling operations*. Let $f_i : D \to D_i$ be continuous functions for all $i \in I$. Their tupling is a function $\langle f_i \rangle_{i\in I} : D \to \Pi_{i\in I} D_i$ which sends each element d of D to $\langle f_i(d) \rangle_{i \in I}$. This is a continuous function as well.

Let $h : D \to \Pi_{i\in I} D_i$. Then h is continuous if and only if for all $i \in I$ the functions $h \circ \pi_i : D \to D_i$ are continuous. Let $f : D_1 \times D_2 \times \cdots \times D_m \to E$ be a function. It is continuous if and only if it is so in each argument separately.

There are two important functions related to product and function space. One is the application, *app*, a continuous function from $[D \to E] \times D$ to E, defined as

$$app(f, x) = f(x).$$

The other is currying. *curry* is a continuous function from $(D \times E) \to F$ to $D \to [E \to F]$, defined as

$$curry(g) = \lambda x \in D.\, (\lambda y \in E.\, g(x, y)).$$

In other words, for any continuous function $g : (D \times E) \to F$, $curry(g)$ is a function which assigns any element x in D a function $g(x, y) : E \to F$.

2.2.3 Sum

Let $\{ D_i \mid i \in I \}$ be an indexed family of cpos. Their *sum*, $\Sigma_{i \in I} D_i$, consists of

$$(\bigcup_{i \in I} \{ i \} \times [D_i \setminus \{ \bot_{D_i} \}]) \cup \{ \bot \}$$

ordered by $x \sqsubseteq y$ if $x = \bot$ or

$$\exists i \in I. \exists d, d' \in D_i. \, d \sqsubseteq d' \ \&\ x = \langle i, d \rangle \ \&\ y = \langle i, d' \rangle.$$

When I is finite we write $D_1 + D_2 + \cdots + D_m$. $\Sigma_{i \in I} D_i$ is obviously a cpo. There are nature functions associated with the sum construction, called *injections*. The j-th injection $in_j : D_j \to \Pi_{i \in I} D_i$ is given by letting $in_j(\bot_{D_j}) = \bot$ and $in_j(d) = \langle j, d \rangle$ if $d \neq \bot_{D_j}$. It is easy to see that injections are continuous and strict.

The sum of two cpos D_1 and D_2 can be pictured as:

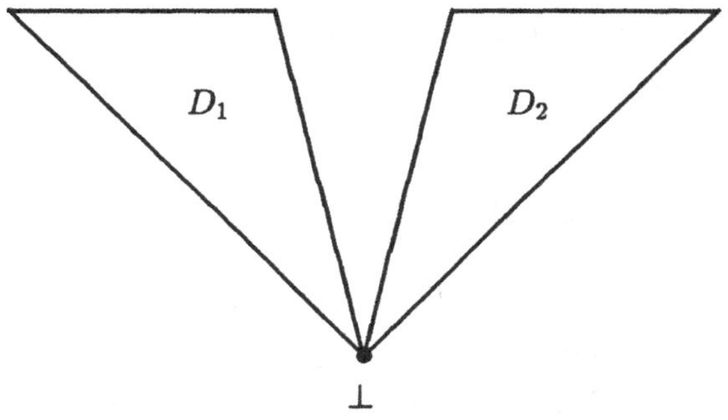

Note that our definition of sum is also referred to as the coaleased or amalgamated sum, different from the separated sum where a new bottom is introduced but the old bottoms are retained, both being bigger than the new one.

Given a family of strict continuous function $f_i : D_i \to E$, $i \in I$, we can extend them to a function $[f_i]_{i \in I} : \Sigma_{i \in I} D_i \to E$ by letting

$$[f_i]_{i \in I}(\bot) = \bot \ \text{ and } \ [f_i]_{i \in I}(\langle k, x \rangle) = f_k(x).$$

$[f_i]_{i \in I}$ is the unique strict continuous function f such that $in_i \circ f = f_i$ for all $i \in I$.

2.2.4 Lifting

Let D be a cpo. Its *lifting* $D_\perp$ consists of the set

$$\{\langle 0, d\rangle \mid d \in D\} \cup \{\perp\}$$

ordered by $x \sqsubseteq y$ if $x = \perp$ or there exist d, d' in D such that $d \sqsubseteq d'$, $x = \langle 0, d\rangle$, and $y = \langle 0, d'\rangle$. In picture, the lift of D looks like

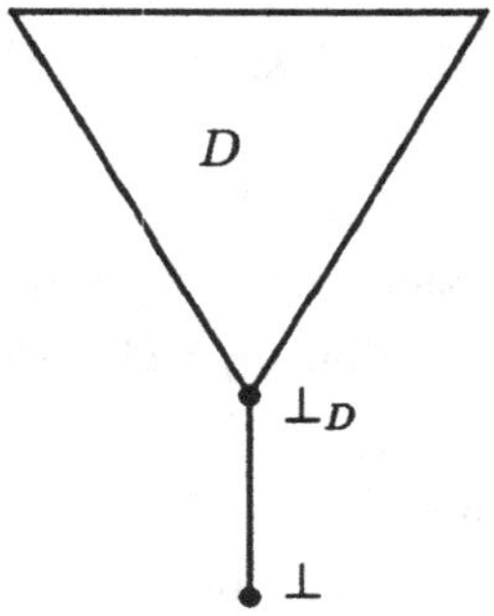

There is a one-one correspondence between strict continuous functions $[D_\perp \to_\perp E]$ and continuous functions $[D \to E]$. The separated sum of a family D_i, $i \in I$ of cpos is the cpo $\Sigma_{i \in I}(D_i)_\perp$.

2.3 Fixed-Point Theory

Let D be a cpo and $f : D \to D$ a monotonic function. An element $a \in D$ is called a *pre-fixed point* of f if $f(a) \sqsubseteq a$; $b \in D$ is called a *post-fixed point* of f if $b \sqsubseteq f(b)$; $c \in D$ is called a *fixed-point* of f if $f(c) = c$. The following theorem is due to Tarski [Ta55].

Theorem 2.3 *(Knaster-Tarski) Let* $(A, \sqsubseteq)$ *be a complete lattice and*

$$f : A \to A$$

a monotonic function. Then

$$\bigsqcap\{a \in A \mid f(a) \sqsubseteq a\}$$

is the least fixed point of f *and*

$$\bigsqcup\{b \in A \mid b \sqsubseteq f(b)\}$$

the greatest fixed point of f.

When it comes to cpos the greatest lower bound of all pre-fixed points is still the least fixed point. We can write this fact in the form of a rule

$$\frac{f(x) \sqsubseteq x}{fix\, f \sqsubseteq x}$$

where $fix\, f$ stands for $\bigsqcap\{a \in D \mid f(a) \sqsubseteq a\}$. It is clear that

$$f\,(fix\, f) = fix\, f.$$

We can view fix again as a function, given by $fix\,(f) = fix\, f$. Such a function $fix : [D \to D] \to D$ is, as expected, continuous.

There is a general method to construct the least fixed-point of a continuous function in any cpo.

Theorem 2.4 *(Kleene) Let D be a cpo, and $f : D \to D$ a continuous function. Then*

$$\bigsqcup_{i \in \omega} f^i\,(\bot)$$

is the least fixed point of f, where $f^0 = \lambda x.\, x$, the identity function, and f^i stands for the composition of f with itself i times.

To reason about least fixed-points there is the Scott induction principle. Let P be a property of elements of a cpo D. P is *inductive* if $P(\bot)$ and for all chain

$$x_0 \sqsubseteq x_1 \sqsubseteq x_2 \cdots \sqsubseteq x_n \cdots$$

$\forall i.\, P(x_i)$ implies $P(\bigsqcup_{i \in \omega} x_i)$.

Theorem 2.5 *(Scott induction) Let D be a cpo, P an inductive property of D, and $f : D \to D$ a continuous function. If*

$$\forall x \in D.\, P(x) \Longrightarrow P(f(x))$$

then $P(fix\ f)$.

2.4 Scott Topology

A topology on a set S is a collection of subsets of S that is closed under finite intersection and arbitrary union. A set S and a topology $\Re$ on it form a topological space $(S, \Re)$. A base of the topology $\Re$ on S is a subset $\Re_1 \subseteq \Re$ such that every open set is the union of elements of $\Re_1$. Let $(D, \sqsubseteq)$ be a cpo. A subset O of D is said to be *Scott open* if O is upwards closed, *i.e.*, $\forall x \in O \forall y \in D$, $x \sqsubseteq y$ implies $y \in O$, and whenever $X \subseteq D$ is directed and $\bigsqcup X \in O$ we have $X \cap O \neq \emptyset$. The *Scott topology* on a cpo $(D, \sqsubseteq)$ consists of all the Scott open sets of $(D, \sqsubseteq)$, written $\Omega(D)$. There is also a definition of Scott open set in terms of increasing chains. Again these two definitions agree on ω-algebraic cpos and can be used interchangeably for ω-algebraic cpos accordingly.

Example 2.5 *Consider the product $\mathcal{T} \times \mathcal{O}$.*

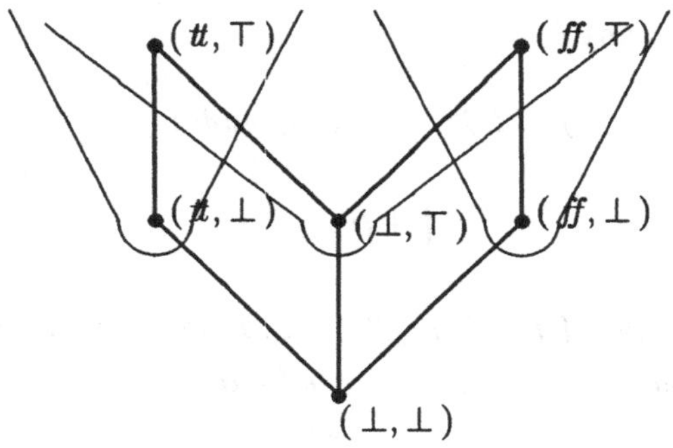

The three 'parabolas' in the picture represent three open sets:

$$\{(tt, \bot), (tt, \top)\},$$
$$\{(\bot, \top), (tt, \top), (ff, \top)\}, \quad and$$
$$\{(ff, \bot), (ff, \top)\}.$$

An open set K of a topological space $(S, \Re)$ is compact provided that any family of open sets whose union contains K has a finite subfamily whose union also contains K. For Part I from now on all domains concerned are supposed to be ω-algebraic and the topology is the Scott topology unless otherwise indicated.

Proposition 2.1 *$\uparrow x$ is open if and only if $x \in D^0$, where*

$$\uparrow x = \{ y \in D \mid x \sqsubseteq y \}.$$

Open sets of this form are called *prime open* since they are the complete primes in the lattice $\Omega(D)$ in the sense that if they are dominated by a join they are dominated by an element of it. In general, we write $\uparrow X$ for the set $\bigcup\{\uparrow x \mid x \in X\}$.

Proposition 2.2 *The set* $\{\uparrow x \mid x \in D^0\}$ *is a base for the Scott topology of an* ω*-algebraic domain* D.

Thus every Scott open set is a union of open sets of the form $\uparrow x$ with $x \in D^0$. The empty set, being open, is the union of the empty collection of prime open sets.

Proposition 2.3 *A non-empty* $O \subseteq D$ *is open and compact if and only if for some finite* $X \subseteq D^0$, $O = \uparrow X$.

The collection of the compact Scott open sets of a domain D is written as $\mathbf{K\Omega}(D)$.

Scott topology characterizes continuous functions. A function f from D to E is continuous if and only if the inverse image of any Scott open set is Scott open. Let $f, g : D \to E$ be continuous functions. $f \sqsubseteq g$ if and only if for every Scott open O of E, $f^{-1}(O) \subseteq g^{-1}(O)$.

2.5 SFP Domains

SFP domains (or **SFP** objects, bifinite domains) are directed limits of Sequence of Finite inductive Partial orders introduced by Plotkin [Pl76] to deal with the denotational semantics of non-deterministic programming languages. A typical phenomenon of SFP domains is that compatible sets of elements need not have least upper bounds (consistently completeness does not hold).

Example 2.6 *An SFP domain which is not a Scott domain:*

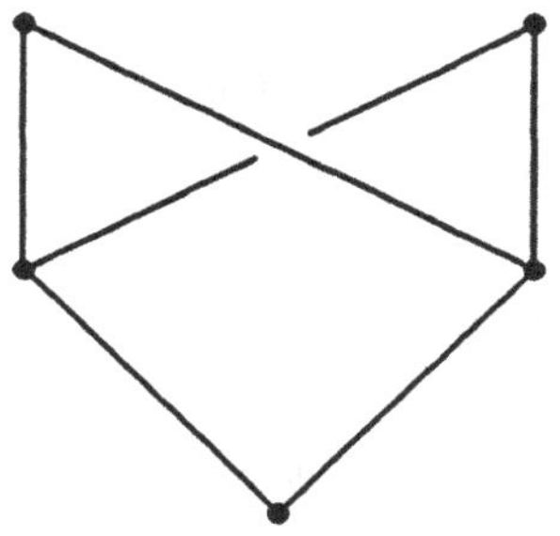

We mention a useful characterization of SFP domains in terms of the minimal upper bounds. Let D be a cpo. A *minimal upper bound* of a subset $X \subseteq D$ is an upper bound of X and it is not strictly greater than any other upper bound of X. Let $\bowtie X$ stand for the set of minimal upper bounds of X. The symbol $\bowtie$ reminds us of the picture in Example 2.6, describing the simplest situation where the minimal upper bound, instead of the least upper bound, is needed. The set $\bowtie X$ is said to be *complete* if whenever u is an upper bound of X, $v \sqsubseteq u$ for some $v \in \bowtie X$. For a subset X of D, let

$$U^0(X) = X,$$
$$U^{i+1}(X) = \{u \mid u \in \bowtie S \ \& \ S \Subset U^i(X)\}, \text{ for } i \geq 0.$$

It is convenient to write $U^*(X)$ for $\bigcup_{i \geq 0} U^i(X)$. Here and from now on, we write $A \Subset B$ to mean A is a finite subset of B.

Theorem 2.6 *(Plotkin [Pl76]) D is an SFP domain if and only if the following three hold:*

1. *D is ω-algebraic,*
2. *$\forall X \Subset D^0$. $\bowtie X$ is complete, and*
3. *$U^*(X)$ is finite.*

The first condition, that D is ω-algebraic, is natural. As far as computation is concerned there seems to be no absolute necessity to go beyond ω-algebraic cpos. The second and third condition, however, are vital. Without one of them the function space cannot preserve ω-algebraicity [Sm83b].

The category **SFP** consists of, as objects, the SFP domains and as morphisms the continuous functions.

Note that if $X \Subset D^0$, every element of $\bowtie X$ is finite, no matter what kind of cpo D is. Let X, Y be subsets of a cpo D. Clearly

$$\uparrow X \cap \uparrow Y = \uparrow \{ \bowtie\{x, y\} \mid x \in X \ \& \ y \in Y\}.$$

From this equality and Theorem 2.6 it follows that, for an SFP domain, finite intersection preserves compact Scott open sets.

2.6 Powerdomains

The purpose of this section is to present the powerdomains of D as subsets of $\mathcal{P}(D)$ in a concise and systematic manner. Special care has been taken here in order to interpret the modal assertions appropriately later.

Let D be an ω-algebraic cpo. There are three popular preorders on the powerset $\mathcal{P}(D)$: $\lesssim_0$, $\lesssim_1$, and $\lesssim_2$. They are defined as

$$A \lesssim_0 B \text{ if } \forall b \in B \exists a \in A.\, a \sqsubseteq b,$$
$$A \lesssim_1 B \text{ if } \forall a \in A \exists b \in B.\, a \sqsubseteq b, \text{ and}$$
$$A \lesssim_2 B \text{ if } A \lesssim_0 B \ \&\ A \lesssim_1 B.$$

Powerdomains can be constructed from these preorders by a technique called *ideal completion*. Let $(P, \lesssim)$ be a preorder with a least element. An ideal of $(P, \lesssim)$ is a subset $X \subseteq P$ which is downwards-closed and directed (hence non-empty). The set of ideals of $(P, \lesssim)$ is written as $I(P)$. $(I(P), \subseteq)$ is an algebraic domain with isolated elements $\{q \in P \mid q \lesssim p\}$ for $p \in P$. $(I(P), \subseteq)$ is called the ideal completion of $(P, \lesssim)$.

The Smyth powerdomain, the Hoare powerdomain and the Plotkin powerdomain of an ω-algebraic cpo D are the ideal completions of $(M[D], \lesssim_0)$, $(M[D], \lesssim_1)$, and $(M[D], \lesssim_2)$, respectively, where $M[D]$ consists of the finite, non-empty, subsets of D^0. One reason for considering the subset $M[D]$ of $\mathcal{P}(D)$ rather than the whole powerset is to make sure the resulting cpos are ω-algebraic again. Without restricting to $M[\mathcal{N}_\perp]$, for example, the ideal completion of $(\mathcal{P}(\mathcal{N}_\perp), \lesssim_2)$ would have more than a countable number of finite elements.

Although conceptually simple, the powerdomains constructed from ideal completions have the disadvantage that they do not really live in the powerset $\mathcal{P}(D)$. The elements of $M[D]$ are already subsets of D, so the ideals are some sets of subsets of D, and are elements of $\mathcal{P}(\mathcal{P}(D))$ instead of $\mathcal{P}(D)$.

In Chapter 3 and Chapter 4, a more concrete presentation of powerdomains of D is needed, so that the elements of the powerdomains are subsets of D. The idea is to pick up the biggest element (with respect to set inclusion) in each equivalent class induced by the three preorders.

For an ω-algebraic cpo D, define, on the powerset $\mathcal{P}(D)$, the following three operations:

$$\mathcal{C}l_S(A) = \uparrow A,$$
$$\mathcal{C}l_H(A) = \downarrow A, \text{ and}$$
$$\mathcal{C}l_P(A) = \mathcal{C}l_S(A) \cap \mathcal{C}l_H(A),$$

where $\downarrow A = \{x \in D \mid \exists a \in A.\, x \sqsubseteq a\}$. Clearly these operations are idempotent, *i.e.*, $\mathcal{C}l(\mathcal{C}l(A)) = \mathcal{C}l(A)$ for every $A \in \mathcal{P}(D)$.

Note that $\mathcal{C}l_S(A) \subseteq \mathcal{C}l_S(B)$ if and only if $B \precsim_0 A$. The Smyth powerdomain of an ω-algebraic domain D consists of, as the finite elements, all subsets $\mathcal{C}l_S(A)$ with A a non-null finite set of D^0. Such a collection of elements form a partial order under the superset ordering, and we have to add to it the limits (least upper bounds) to get a cpo. The limit point of an ω-increasing chain

$$\mathcal{C}l_S(A_0) \precsim_0 \mathcal{C}l_S(A_1) \cdots \precsim_0 \mathcal{C}l_S(A_i) \precsim_0 \cdots$$

is the intersection $\bigcap_{i\in\omega} \mathcal{C}l_S(A_i)$. This intersection must be non-empty provided that $A_i \in M[D]$ for every i. Let $A_0 = \{ x_0, x_1, \cdots, x_k \}$. It is easy to see that for some j, $0 \leq j \leq k$, we have

$$\uparrow x_j \cap \mathcal{C}l_S(A_i) \neq \emptyset$$

for every i. That implies $\bigsqcup(\uparrow x_j) \in \mathcal{C}l_S(A_i)$ for every i. This result can be extended to the general situation where each $\mathcal{C}l_S(A_i)$ is the limit of an ω-increasing chain.

Let $\mathcal{P}_S(D)$ be the set consisting of $\mathcal{C}l_S(A)$ with $A \in M[D]$ together with the limits of the ω-increasing chains of such elements. We have

Theorem 2.7 *Let D be an ω-algebraic cpo. Then $(\mathcal{P}_S(D), \precsim_0)$ is the Smyth powerdomain of D. It is an ω-algebraic cpo isomorphic to the ideal completion of the preorder $(M[D], \precsim_0)$.*

Proof Every element of $\mathcal{P}_S(D)$ is a fixed-point of $\mathcal{C}l_S$. Hence

$$(\mathcal{P}_S(D), \precsim_0)$$

is a partial order. D is its bottom.

It is easy to see that the intersection $\bigcap_i \mathcal{C}l_S(A_i)$ is the least upper bound of an ω-increasing chain

$$\mathcal{C}l_S(A_0) \precsim_0 \mathcal{C}l_S(A_1) \cdots \precsim_0 \mathcal{C}l_S(A_i) \precsim_0 \cdots.$$

By a fairly elementary argument one can show that $\bigcap_{i\in\omega} \mathcal{C}l_S(A_i)$ is still in $\mathcal{P}_S(D)$ as long as the ω-increasing chain is from it. That means

$$(\mathcal{P}_S(D), \precsim_0)$$

is a cpo. The ω-algebracity of $(\mathcal{P}_S(D), \sqsubseteq_0)$ follows if it can be shown that every $\mathcal{C}l_S(A)$ is a finite element when A is from $M[D]$, by virtue of Lemma 2.1. Assume it were not the case. There would be a situation such that for some chain

$$\mathcal{C}l_S(B_0) \sqsubseteq_0 \mathcal{C}l_S(B_1) \cdots \sqsubseteq_0 \mathcal{C}l_S(B_i) \sqsubseteq_0 \cdots$$

we have

$$\bigcap_{k\in\omega} \mathcal{C}l_S(B_k) \subseteq \mathcal{C}l_S(A)$$

but $\mathcal{C}l_S(B_k) \not\subseteq \mathcal{C}l_S(A)$ for any k, where $A \in M[D]$ and $B_k \in M[D]$ for all k. In other words, for any k, there is some $b_k \in B_k$ such that $b_k \not\sqsupseteq_D a$ for every $a \in A$. Since $\mathcal{C}l_S(B_i) \supseteq \mathcal{C}l_S(B_j)$ when $i \leq j$, for each k there is a finite decreasing chain

$$b_k \sqsupseteq_D c_{k-1} \sqsupseteq_D c_{k-2} \cdots \sqsupseteq_D c_0$$

such that $c_i \in B_i$ and $c_i \not\sqsupseteq_D a$ for any $a \in A$, $0 \leq i < k$. All B_k's are finite set. Therefore there exists a finite branching, infinite tree with the root in B_0 and each node being an element of some B_k. By König's lemma, this tree has an infinite branch

$$d_0 \sqsubseteq_D d_1 \sqsubseteq_D \cdots \sqsubseteq_D d_i \cdots.$$

This determines an element $\bigsqcup_{i\in\omega} d_i$ of $\bigcap_{k\in\omega} \mathcal{C}l_S(B_k)$, which leads to a contradiction.

A one-one, order preserving correspondence between $(\mathcal{P}_S(D), \sqsubseteq_0)$ and the ideal completion of $(M[D], \sqsubseteq_0)$ can now be easily established. This justifies our construction of the Smyth powerdomain.

□

For any $A \in \mathcal{P}(D)$, consider the partial order

$$(\{X \mid X \equiv_0 A\}, \subseteq),$$

where $\equiv_0$ is the equivalent relation induced by $\sqsubseteq_0$. It is a fact that T is the biggest in this partial order if and only if $\mathcal{C}l_S(T) = T$. Therefore, all the elements in the Smyth powerdomain are the biggest in their equivalent classes.

Similarly, $\mathcal{C}l_H(A) \subseteq \mathcal{C}l_H(B)$ if and only if $A \sqsubseteq_1 B$, and the Hoare powerdomain of an ω-algebraic domain D consists of, as finite elements, all subsets $\mathit{Cl}_H(A)$ with A a non-null finite set of D^0. Let $\mathcal{P}_H(D)$ stand for the collection of such elements together with their limits described below. The limit point of an ω-increasing chain

$$\mathit{Cl}_H(A_0) \subseteq \mathit{Cl}_H(A_1) \cdots \subseteq \mathit{Cl}_H(A_i) \subseteq \cdots$$

is the union

$$\bigcup_{i \in \omega} \mathit{Cl}_H(A_i).$$

Theorem 2.8 *For an ω-algebraic cpo D, $(\mathcal{P}_H(D), \sqsubseteq_1)$ is the Hoare powerdomain of D. It is an ω-algebraic cpo isomorphic to the ideal completion of the preorder $(M[D], \sqsubseteq_1)$.*

The proof is similar to that of the previous theorem, hence omitted. Note all the members of the Hoare powerdomain are the largest (under inclusion) elements in their equivalent classes.

It can be shown that

$$\begin{array}{l} \mathit{Cl}_P(A) \sqsubseteq_2 \mathit{Cl}_P(B) \iff A \sqsubseteq_2 B, \text{ and} \\ A \sqsubseteq_2 B \ \&\ B \sqsubseteq_2 A \iff \mathit{Cl}_P(A) = \mathit{Cl}_P(B). \end{array}$$

However, the Plotkin powerdomain of an ω-algebraic cpo D is slightly complicated. It consists of, as finite elements, all subsets $\mathit{Cl}_P(A)$ with A a non-empty finite set of D^0. The order is $\sqsubseteq_2$. The limit point of an ω-increasing chain

$$\mathit{Cl}_P(A_0) \sqsubseteq_2 \mathit{Cl}_P(A_1) \cdots \sqsubseteq_2 \mathit{Cl}_P(A_i) \sqsubseteq_2 \cdots$$

is defined to be the closure $\mathit{Cl}_P(A)$ where A consists of the least upper bounds of increasing chains

$$a_0 \sqsubseteq a_1 \sqsubseteq a_2 \cdots \sqsubseteq a_n \sqsubseteq \cdots$$

such that $\forall i \in \omega.\ a_i \in A_i$. Incidentally, least upper bounds of ω-increasing chains in the Smyth and the Hoare powerdomain can be defined in a similar manner (see the proof of Theorem 2.9).

Write $\mathcal{P}_P(D)$ for the set consisting of closures $\mathit{Cl}_P(B)$ with $B \in M[D]$ together with all the limits of ω-increasing chains given above. It has to be shown that the limit $\mathit{Cl}_P(A)$ is indeed the least upper bound with respect to $\sqsubseteq_2$ in $\mathcal{P}_P(D)$ (note that it is not possible to prove that the limit is the least upper bound in the powerset $\mathcal{P}(D)$). The following theorem justifies the construction of the Plotkin powerdomain. In particular, the limit construction gives least upper bounds.

Theorem 2.9 *Given an ω-algebraic cpo D, $(\mathcal{P}_P(D), \sqsubseteq_2)$ is the Plotkin powerdomain. It is an ω-algebraic cpo isomorphic to the ideal completion of the preorder $(M[D], \lesssim_2)$.*

Proof It is enough to show that $\mathcal{P}_P(D)$ is a cpo with all the finite elements being of the form $Cl_P(A)$, with $A \in M[D]$. It is easy to justify that $\mathcal{P}_P(D)$ is a partial order since for any $B \in \mathcal{P}_P(D)$, $Cl_P(B) = B$. It is obvious that $\{ \bot_D \}$ is the bottom element of $\mathcal{P}_P(D)$.

Suppose

$$Cl_P(A_0) \lesssim_2 Cl_P(A_1) \cdots \lesssim_2 Cl_P(A_i) \lesssim_2 \cdots$$

is a chain in $\mathcal{P}_P(D)$. Let A be the set consisting of the least upper bounds of ω-increasing chains

$$a_0 \sqsubseteq a_1 \sqsubseteq a_2 \cdots \sqsubseteq a_n \sqsubseteq \cdots$$

such that $\forall i \in \omega.\ a_i \in A_i$. It is easy to show that $Cl_H(A) = \bigcup_{j \in \omega} Cl_H(A_j)$. Similarly, one can prove that $Cl_S(A) = \bigcap_{i \in \omega} Cl_S(A_i)$ by using König's lemma. By Theorem 2.7, Theorem 2.8, and their proofs, it is then an easy step to check that $Cl_P(A)$ is the least upper bound, which equals to the limit of a similar chain but with every A_i in $M[D]$. Therefore $(\mathcal{P}_P(D), \lesssim_2)$ is a complete partial order.

It remains to show that $Cl_P(A)$ with $A \in M[D]$ are finite elements. Suppose $A \in M[D]$ and $Cl_P(A) \lesssim_2 Cl_P(B)$, where $Cl_P(B)$ is the limit of the chain

$$Cl_P(B_0) \lesssim_2 Cl_P(B_1) \cdots \lesssim_2 Cl_P(B_i) \lesssim_2 \cdots$$

with $B_i \in M[D]$ for each i. A is a finite set of finite elements. From the definition of B it can be seen that there is some n, $A \lesssim_1 B_n$. By a similar argument used in the proof of Theorem 2.7, it can be shown that for some $m > n$, $A \lesssim_0 B_m$. Therefore $A \lesssim_2 B_m$ by the transitivity of $\lesssim_1$.

□

We leave it to the reader to check that the elements of the Plotkin powerdomain are maximal in their equivalent classes with respect to inclusion.

2.7 Finite Elements

This section gives a description of finite elements with respect to constructions on domains. The first proposition specifies the finite elements in the product, the sum, and the lifting. The proof is straightforward.

Proposition 2.4 *Let D and E be ω-algebraic cpos and D^0, E^0 the set of finite elements of D and E, respectively. We have*

$$
\begin{aligned}
(D \times E)^0 &= \{ \langle d, e \rangle \mid d \in D^0, e \in E^0 \}, \\
(D + E)^0 &= \{ \langle 0, d \rangle \mid d \in D^0 \ \&\ d \neq \bot_D \} \cup \\
&\qquad \{ \langle 1, e \rangle \mid e \in E^0 \ \&\ e \neq \bot_E \} \cup \{ \bot \}, \\
(D_\bot)^0 &= \{ \langle 0, d \rangle \mid d \in D^0 \} \cup \{ \bot \}.
\end{aligned}
$$

To deal with finite elements in the function space, some definitions are required. Let D, E be cpos. A one-step function is a function $[a, b]$ defined as

$$
[a, b](x) = \begin{cases} b & \text{if } x \sqsupseteq a , \\ \bot & \text{otherwise} \end{cases}
$$

where $a \in D^0$, $b \in E^0$. This is very much like the characteristic function of the open set $\uparrow a$. For Scott domains D and E, $f : D \to E$ is called a *step function* if

$$
f = \bigsqcup_{i \in I} [a_i, b_i]
$$

for some $a_i \in D^0$, $b_i \in E^0$, $i \in I$, with I finite such that

$$
\forall J \subseteq I.\ \{ a_j \mid j \in J \} \uparrow \Longrightarrow \{ b_j \mid j \in J \} \uparrow .
$$

Note under this condition,

$$
\bigsqcup_{i \in I} [a_i, b_i] = \lambda x.\ \bigsqcup \{ b_i \mid a_i \sqsubseteq x \ \&\ i \in I \}.
$$

Step functions represent exactly all the finite elements in the function space. We have

Proposition 2.5 *Let D, E be Scott domains.*

$$
[D \to E]^0 = \{ f \mid f \ \textit{is a step function of}\ [D \to E] \}.
$$

The proof for this proposition is not difficult. Finite elements in the function space of SFP domains are harder to describe, however. They are treated in Section 2.10.

The following proposition is a corollary of the theorems given in the previous section.

Proposition 2.6 *If D is an ω-algebraic cpo then*

$$(\mathcal{P}_S[D])^0 = \{ \mathcal{Cl}_S(A) \mid A \in M[D] \},$$
$$(\mathcal{P}_H[D])^0 = \{ \mathcal{Cl}_H(A) \mid A \in M[D] \}, \quad \textit{and}$$
$$(\mathcal{P}_P[D])^0 = \{ \mathcal{Cl}_P(A) \mid A \in M[D] \}.$$

2.8 Compact Open Sets

A compact open set can be regarded as a property about processes by finite observation [HeMi79]. In this section we show that all the constructions introduced so far preserve compactness. It provides a basis so that properties of composite types can be built up from those of simpler types.

Remember that the collection of Scott open sets of of a cpo D is written as $\Omega(D)$. The compact open sets of $\Omega(D)$ are exactly the finite elements of the lattice $(\Omega(D), \subseteq)$. Clearly D and $\emptyset$ are always compact open. For SFP domains, the union and intersection of two compact open sets remain to be compact open. Moreover, product, sum, and lifting preserve compactness.

Proposition 2.7 *Let A, B be compact open sets in $\Omega(D)$ and $\Omega(E)$, respectively, where D, E are SFP domains. Then*

$\{ \langle u, v \rangle \mid u \in A \;\&\; v \in B \}$ *is a compact open set of $D \times E$,*
$\{ \langle 0, u \rangle \mid u \in A \}$, $\{ \langle 1, v \rangle \mid v \in B \}$ *are compact open sets of $D + E$*
provided that $\bot_D \notin A$, $\bot_E \notin B$, and
$\{ \langle 0, u \rangle \mid u \in A \}$ *is a compact open set of $D_\bot$.*

Note that if $\bot_D \in A$ or $\bot_E \in B$ then $A = B = (D + E)$. The case for function space is more interesting.

Proposition 2.8 *If A and B are compact open sets of $\Omega(D)$ and $\Omega(E)$, respectively, where D, E are SFP domains, then*

$$A \to B = \{ f \in [D \to E] \mid A \subseteq f^{-1}(B) \}$$

is a compact open set of $[D \to E]$.

Proof We have, for $a \in D^0$ and $b \in E^0$,

$$\begin{aligned} \uparrow a \to \uparrow b &= \{ f \in [D \to E] \mid \uparrow a \subseteq f^{-1}(\uparrow b) \} \\ &= \{ f \in [D \to E] \mid f(a) \sqsupseteq b \} \\ &= \uparrow [a, b]. \end{aligned}$$

Therefore, $\uparrow a \to \uparrow b$ is a compact open set.

By Proposition 2.3, A and B can be written as $A = \uparrow X$ and $B = \uparrow Y$, with X, Y finite subsets of D^0. It is not difficult to see that

$$\begin{aligned} A \to B &= \uparrow X \to \uparrow Y \\ &= \bigcap_{a \in X} (\uparrow a \to \uparrow Y) \\ &= \bigcap_{a \in X} (\bigcup_{b \in Y} \uparrow a \to \uparrow b) \\ &= \bigcap_{a \in X} (\bigcup_{b \in Y} \uparrow [a, b]). \end{aligned}$$

Hence $A \to B$ is compact open as finite union and intersection of compact open sets are compact open.

□

Powerdomain constructions also preserve compactness.

Proposition 2.9 *Let* $A = \uparrow X$ *be a compact open set in* $\Omega(D)$, *with* D *an SFP domain, and* X *a finite set of isolated elements. We have*

$$\{ U \in \mathcal{P}_S(D) \mid U \subseteq A \} = \uparrow \mathcal{C}l_S(X)$$

and

$$\{ L \in \mathcal{P}_H(D) \mid L \cap A \neq \emptyset \} = \bigcup_{a \in X} \uparrow \mathcal{C}l_H\{a\}.$$

This result implies that

$$\{ U \in \mathcal{P}_S(D) \mid U \subseteq A \}$$

is a compact open set of the Smyth powerdomain $\mathcal{P}_S(D)$, and

$$\{ L \in \mathcal{P}_H(D) \mid L \cap A \neq \emptyset \}$$

a compact open set of the Hoare powerdomain $\mathcal{P}_H(D)$.

The same construction can produce compact open sets in a the Plotkin powerdomain.

Proposition 2.10 *If A is a compact open set in $\Omega(D)$, with D an SFP domain, then*

$$\{ U \in \mathcal{P}_P(D) \mid U \subseteq A \}$$

and

$$\{ L \in \mathcal{P}_P(D) \mid L \cap A \neq \emptyset \}$$

are compact open sets of $\mathcal{P}_P(D)$.

Proof When A is empty there is nothing to prove. Let

$$A = \uparrow X$$

where X is a finite set of pairwise incomparable, isolated elements. For any U in $\mathcal{P}_P(D)$, we have

$$\begin{aligned} U \subseteq \uparrow X &\iff \forall x \in U\ \exists a \in X.\ x \sqsupseteq a \\ &\iff \exists Y \subseteq X.(\forall x \in U\ \exists a \in Y.\ a \sqsubseteq x\ \&\forall b \in Y\ \exists x \in U.\ b \sqsubseteq x) \\ &\iff \exists Y \subseteq X. Y \sqsubseteq_2 U. \end{aligned}$$

Note that Y must be non-empty. Hence

$$\{ U \in \mathcal{P}_P(D) \mid U \subseteq A \} = \bigcup_{Y \subseteq X\ \&\ Y \neq \emptyset} \uparrow Cl_P(Y),$$

a compact open set.

Similarly,

$$\{ L \in \mathcal{P}_P(D) \mid L \cap A \neq \emptyset \} = \bigcup_{a \in X} \uparrow Cl_P(\{a, \bot_D\}).$$

Hence $\{ L \in \mathcal{P}_P(D) \mid L \cap A \neq \emptyset \}$ is also compact open.

□

We remark that a stronger result holds for all the constructions, that is, not only compact open sets are preserved by the constructions, but prime opens are also preserved (for the Plotkin powerdomain we have to use the intersection). The proofs are extremely similar.

2.9 Recursively Defined Domains

For the purpose of denotational semantics, recursively defined domains are often needed. The subject started with Scott's pioneer work on the solution of the equation $D \cong [D \to D]$ using an inverse limit construction. Such a solution lay a formal basis for the untyped λ-calculus, where a term is sometimes used as an object, sometimes as a function. Scott's work was later put into a category-theoretic setting by Smyth and Plotkin [SmPl82].

The same idea is used in Larsen and Winskel's work [LaWi84] to give recursively defined domains in the category of information systems. Information systems are a representation of Scott domains. The fact that they are based concretely on sets and relations makes it possible to introduce a sub-structure relation on them to form a cpo. All the usual domain constructions have their counterparts as continuous operations on information systems. Recursive domain equations can then be solved by Kleene's construction of least fixed-points. The construction of the fixed-point makes the domain isomorphism an equality.

In later chapters we will present some work following the style of Larsen and Winskel. Therefore, no further detail is provided in this section.

2.10 More about SFP Domains

We first introduce a new characterization of finite elements in the function space of SFP domains. We adopt this characterization not only because it is simple(r), but also because the same characterization is used later in defining the approximable mappings on strongly finite sequent structures, as well as in capturing the prime assertions in the logic of SFP.

Let D, E be SFP domains and D^0, E^0 the set of finite elements of D and E, respectively. How can the finite elements of $[D \to E]$ be characterized? We know that for Scott domains the finite elements of the function space are step functions

$$\bigsqcup_{i \in I} [\, a_i,\, b_i \,]$$

where I is a finite set and $a_i \in D^0$, $b_i \in E^0$ are such that

$$\forall J \subseteq I.\ \{\, a_j \mid j \in J \,\}\uparrow \Longrightarrow \{\, b_j \mid j \in J \,\}\uparrow .$$

For SFP domains, however, more information is needed to get a finite element in the function space. Although each one-step function still specifies a finite element, it is not true that the finite elements of the function space of SFP domains take the same form as those of the function space of Scott domains.

Consider the function space of the following SFP domain to itself.

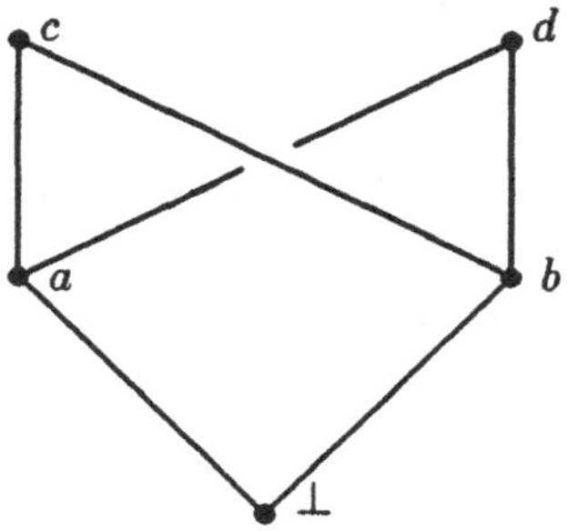

Note $[a, a] \sqcup [b, b]$ is a proper form of step function for Scott domains. But it does not give a well-defined function because it does not tell where c and d should be sent to. For this reason we give the following definition of step functions for SFP domains.

Definition 2.1 *Let D, E be SFP domains. A set $T \Subset D^0 \times E^0$ is called joinable if*

$$\forall T' \subseteq T.\ [\forall a \in \bowtie \pi_1 T' \exists b \in \bowtie \pi_2 T'.\ (a, b) \in T].$$

A function $f : D \to E$ is a step function if there is a joinable set

$$\{ (a_i, b_i) \mid i \in I \}$$

such that $f = \bigsqcup_{i \in I} [a_i, b_i]$.

Note that $T \Subset D^0 \times E^0$ is joinable if

$$\begin{aligned} \forall T' \subseteq T \exists X \subseteq \pi_1 T\ \exists Y. \quad & \bowtie \pi_1 T' = X\ \& \\ & \bowtie \pi_2 T' = Y\ \& \\ & \forall a \in X \exists b \in Y.\ (a, b) \in T. \end{aligned}$$

Putting it in yet another slightly different form, $T \Subset D^0 \times E^0$ is joinable if

$$\forall T' \subseteq T \forall X \forall Y.$$
$$(\bowtie \pi_1 T' = X\ \&\ \bowtie \pi_2 T' = Y) \Longrightarrow (\forall a \in X \exists b \in Y.\ (a, b) \in T).$$

These are equivalent descriptions of joinable sets since the set of minimal upper bounds is unique. However under other contexts they will be different, and we have to make a choice which description to use.

From the definition it can be seen that if

$$\bigsqcup_{i \in I} [\, a_i,\, b_i \,]$$

is a step function then

$$\forall J \subseteq I.\ \{\, a_j \mid j \in J \,\}\!\uparrow \Longrightarrow \{\, b_j \mid j \in J \,\}\!\uparrow .$$

Hence the step functions given here are indeed step functions if restricted to Scott domains. Moreover, they represent all finite elements of the function space for SFP domains. That follows from a result of Gunter (Theorem 2.10) and a proposition below.

Definition 2.2 *(Gunter) Let D, E be SFP domains and*

$$\{\, (\, a_i,\, b_i \,) \mid i \in I \,\} \Subset D^0 \times E^0$$

a finite set. $\{\, (\, a_i,\, b_i \,) \mid i \in I \,\}$ is called Gunter joinable if for any $a \in D^0$,

$$\{\, (\, a_i,\, b_i \,) \mid i \in I \;\&\; a_i \sqsubseteq a \,\}$$

has a maximum in $D^0 \times E^0$.

Theorem 2.10 *(Gunter [Gu87]) Let D, E be SFP and*

$$\{\, (\, a_i,\, b_i \,) \mid i \in I \,\}$$

Gunter joinable. Then

$$\bigsqcup_{i \in I} [\, a_i,\, b_i \,]$$

is a finite element in $[D \to E]$. Moreover functions of this form represent all the finite elements in $[D \to E]$.

The relationship between joinable and Gunter joinable sets is given by the following proposition.

Proposition 2.11 *Let D, E be SFP. If $\{\, (\, a_i,\, b_i \,) \mid i \in I \,\}$ is joinable then it is Gunter joinable.*

Proof Suppose

$$\{ (a_i, b_i) \mid i \in I \}$$

is joinable. Then $\{ a_i \mid i \in I \}$ is $\bowtie$-closed. Let $a \in D^0$. To show

$$Q = \{ (a_i, b_i) \mid i \in I \ \& \ a_i \sqsubseteq a \}$$

has a maximum in $D^0 \times E^0$ let us prove that

$$\{ a_i \mid i \in I \ \& \ a_i \sqsubseteq a \}$$

has a maximum. Suppose

$$S = \bowtie \{ a_i \mid i \in I \ \& \ a_i \sqsubseteq a \}.$$

We have $S \subseteq \{ a_i \mid i \in I \}$ and for some $a' \in S$. $a' \sqsubseteq a$ since S is complete. This a' is clearly the maximum. We know that there is a set finite set T such that $T = \bowtie \pi_2 Q$. Therefore for some $b' \in T$, $(a', b') \in Q$. This (a', b') is clearly the maximal in Q.

□

We remark that the converse of the proposition is not true. Consider the function space from the diamond $(\mathcal{O} \times \mathcal{O})$ to the diamond with a new top t. The set

$$\{((\top, \top), t), ((\top, \bot), (\top, \bot)), ((\bot, \top), (\bot, \top))\}$$

is Gunter joinable but not joinable.

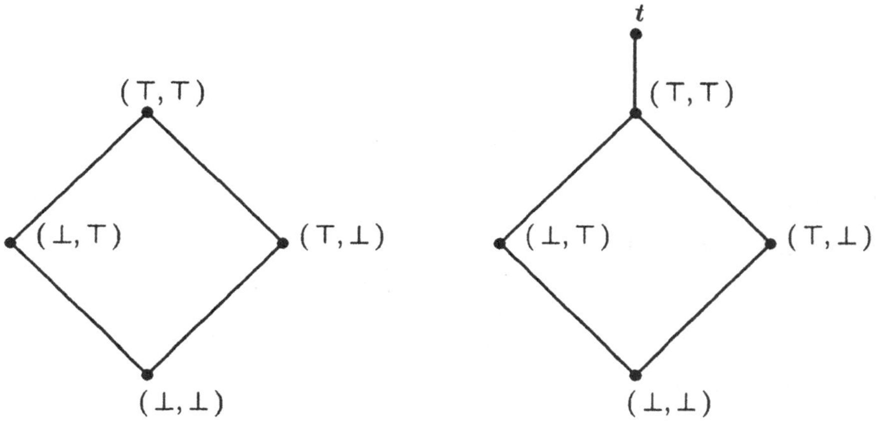

We have, however,

Proposition 2.12 *Let D, E be SFP domains. If G is Gunter joinable then there is a super set Q which is joinable and G, Q defines the same function.*

Proof Suppose G is Gunter joinable. Let $G' \subseteq G$ and $a \in \bowtie \pi_1 G'$. Then

$$F = \{ (x, y) \mid (x, y) \in G \,\&\, x \sqsubseteq a \}$$

has a maximal (a, b), since G is Gunter joinable. Clearly $G' \subseteq F$. Therefore b is an upper bound of $\pi_2 G'$. This implies for some $c \in \bowtie \pi_2 G'$, $c \sqsubseteq b$. It is easy to see that the union $G \cup \{(a, c)\}$ defines the same function as G does. Continuing this process a finite number of times we will get a joinable set. Only finitely many such steps is needed because $\pi_2 G$ has a $\bowtie$-closed supper set, given that E is an SFP domain.

□

Proposition 2.12 implies that

Theorem 2.11 *Let D, E be SFP domains. A function is a finite element in $[D \to E]$ if and only if it is equal to some step function.*

The rest of the section is about a useful characterization of minimal upper bounds.

Proposition 2.13 *Let A, B be finite sets of isolated elements of an ω-algebraic domain D. $\bowtie A = B$ and B is complete if and only if*

$$\bigcap_{a \in A} \uparrow a = \bigcup_{b \in B} \uparrow b \text{ and } \forall b, b' \in B. \, (b \sqsubseteq b' \Rightarrow b = b').$$

Proof Only if: Let $\bowtie A = B$ and B be complete. Clearly

$$\forall b, b' \in B. \, b \sqsubseteq b' \Rightarrow b = b'.$$

Since B is complete, for each upper bound u of A there is some $b \in B$ such that $u \sqsupseteq b$. Therefore

$$\bigcap_{a \in A} \uparrow a \subseteq \bigcup_{b \in B} \uparrow b.$$

The other direction of the inclusion is obvious.

If: Assume $\bigcap_{a \in A} \uparrow a = \bigcup_{b \in B} \uparrow b$ and

$$\forall b, b' \in B. \, b \sqsubseteq b' \Rightarrow b = b'.$$

Suppose m is a minimal upper bound of A. Then it is an upper bound of A. Therefore $m \in \bigcap_{a \in A} \uparrow a$. By the provided equation, $m \in \bigcup_{b \in B} \uparrow b$. Hence $m \sqsupseteq b$ for some $b \in B$. As each member of B is clearly an upper bound of A, we must have $m = b$. So, $\bowtie A \subseteq B$.

Suppose $b \in B$. Clearly it is an upper bound of A. Assume $b \sqsupseteq q$, where q is an upper bound of A. We have $q \in \bigcap_{a \in A} \uparrow a$, which implies the existence of some $b' \in B$ such that $q \sqsupseteq b'$, by the given condition. Therefore $b \sqsupseteq b'$. Hence $b = b'$, also $b = q$. This means $b \in \bowtie A$.

Therefore $\bowtie A = B$. It is easy to see that B is complete.

□

Proposition 2.13 is due to Gunter [Gu87]. One should be careful that the requirement 'B is complete' in the proposition cannot be dropped. The topological counterpart of this result is expressed as the *quasi-conjunctive* property. A set P of prime open sets of domain D is quasi-conjunctive if for all non-null finite subset S of P there is a subset R of P such that

$$\bigcap S = \bigcup R.$$

Chapter 3

A Representation of SFP

This chapter introduces a representation of the category of SFP domains. It uses structures similar to but more general than Scott's information systems. Distilled from Gentzen's sequent calculi, a basic structure called a sequent structure is produced. Sequent structures determine a major part of the axioms of information systems. A category of special kind of sequent structures called the strongly finite ones is shown to be equivalent to the category of SFP domains. Constructions like the Plotkin powerdomain and the function space are given, as well as a complete partial order of such structures to give solutions to recursively defined systems.

As stated in Section 1.3, information systems give a logical approach to domain theory. Therefore the representation is seen as an intermediate step towards the logic of SFP domains, although the results are interesting in themselves.

3.1 Sequent Structures

The sequent calculi due to Gentzen include a proof system for propositional logic. In this proof system there are two logical rules for each logical connective: the introduction rule and the elimination rule. The structural rules of Gentzen's sequent calculi include the identity axiom, the weakening rule, and the cut rule (however, the cut rule has been shown to be redundant by a theorem called cut-elimination).

Without using any logical connectives (and, of course, neither the corresponding logical rules), the sequent calculus becomes a structure which we call 'a sequent structure', abbreviated as SS.

Definition 3.1 *A sequent structure (SS) is a pair* $(A, \vdash)$*, where* A *is a set of propositions, and* $\vdash$ *is an entailment relation satisfying the following rules:*

$$(\textit{Identity}) \qquad a \vdash a,$$

$$(\textit{Weakening}) \qquad \frac{X' \supseteq X \quad X \vdash Y \quad Y \subseteq Y'}{X' \vdash Y'},$$

$$(\textit{Cut}) \qquad \frac{X \vdash Y, a \quad a, X' \vdash Y'}{X, X' \vdash Y, Y'}.$$

To make some sense of a formula $X \vdash Y$, it is important to attach some informal meaning to it. One should read $X \vdash Y$ as *conjunction of* X*'s entails disjunction of* Y*'s.* The notations used here follow standard convention. However, when we write $X \vdash Y$, we not only require X and Y be finite, but also X be non-empty. In addition, $X, X' \vdash Y, Y'$ stands for $X \cup X' \vdash Y \cup Y'$. For convenience, a singleton set $\{a\}$ is written just as a in this context.

Note that in the absence of logical connectives and their related rules, Cut can not be eliminated. Actually Cut is a rather powerful rule, as demonstrated by the following theorem. It would not be true without Cut.

Theorem 3.1 *Let* $(A, \vdash)$ *be a sequent structure. Then*

1. $a \in X \Longrightarrow X \vdash a$,

   $a \in X \Longrightarrow a \vdash X$,
2. $((\forall b \in Y. X \vdash b) \;\&\; Y \vdash Z) \Longrightarrow X \vdash Z$,
3. $(X \vdash Y \;\&\; Y' \subseteq Y \;\&\; \forall b \in Y'. b \vdash Z) \Longrightarrow X \vdash (Y \setminus Y'), Z$,
4. $(\forall b \in Y. X, b \vdash \emptyset) \;\&\; X \vdash Y \Longrightarrow X \vdash \emptyset$,
5. $(\forall b \in Y. X \vdash b) \;\&\; X, Y \vdash \emptyset \Longrightarrow X \vdash \emptyset$.

Proof The first conclusion is trivial.

The second can be shown by repeated applications of the cut rule. When Y is a singleton the conclusion follows by an application of the Cut. In general, let $b \in Y$. We have $X \vdash b$ and $b, (Y \setminus \{b\}) \vdash Z$. Applying the cut rule we get $X, (Y \setminus \{b\}) \vdash Z$. Now choose a $b' \in (Y \setminus \{b\})$, if possible. We have $X \vdash b'$ and $b', (Y \setminus \{b, b'\}), X \vdash Z$. Applying the cut rule again, we get $X, (Y \setminus \{b, b'\}) \vdash Z$. Repeating this process a number of times until $Y \setminus \{b, b', b'', \cdots\}$ is empty, we get $X \vdash Z$.

To prove the third conclusion let $b \in Y'$ (if Y' is empty, then $X \vdash Y, Z$ follows from the Weakening rule). We have $X \vdash (Y \setminus \{b\}), b$ and $b \vdash Z$. It follows from Cut that $X \vdash (Y \setminus \{b\}), Z$. Taking $b' \in (Y' \setminus \{b\})$ and repeating the application of Cut, we get $X \vdash (Y \setminus \{b, b'\}), Z$. After a number of such applications the desired entailment $X \vdash (Y \setminus Y'), Z$ follows.

Since the proofs for the fourth and fifth conclusion are similar, we give a proof for the fourth conclusion as an illustration. The given condition provides

$$X \vdash (Y \setminus \{b\}), b$$

and

$$b, X \vdash \emptyset$$

for all b in Y. We have $X \vdash Y \setminus \{b\}$ as the result of an application of the cut rule. We can repeat this process and finally get $X \vdash \emptyset$.

□

A sequent structure is an axiomatization of the entailment. More formally, one can interpret the propositions as sets, and the entailment $X \vdash Y$ as the relation $\bigcap X \subseteq \bigcup Y$. Under this interpretation it is clear that Identity and Weakening are sound. It can be seen that Cut is also sound. Suppose

$$\bigcap X \subseteq (\bigcup Y) \cup a$$

and

$$a \cap (\bigcap X') \subseteq \bigcup Y'.$$

Given any element x in $(\bigcap X) \cap (\bigcap X')$, we have $x \in \bigcap X \subseteq (\bigcup Y) \cup a$. Either $x \in (\bigcup Y)$, which implies $x \in (\bigcup Y) \cup (\bigcup Y')$, or $x \in a$, which implies $x \in a \cap (\bigcap X')$ and hence $x \in \bigcup Y'$. Therefore,

$$(\bigcap X) \cap (\bigcap X') \subseteq (\bigcup Y) \cup (\bigcup Y').$$

The restriction of the entailment $X \vdash Y$ to non-empty X's is not crucial. It has the effect that any single proposition in a sequent structure would not be too 'weak'. The technical advantage of the restriction will be seen later.

A sequent structure determines a family of subsets of propositions called its elements.

Definition 3.2 *The elements* $|\underline{A}|$, *of a sequent structure* $\underline{A} = (A, \vdash)$ *consists of subsets* x *of propositions which are closed under entailment:*

$$(X \subseteq x \;\&\; X \vdash Y) \Longrightarrow x \cap Y \neq \emptyset.$$

It follows that for x to be an element of $|\underline{A}|$, it must be the case that for no $X \subseteq x$, $X \vdash \emptyset$.

Theorem 3.2 *For a sequent structure* $\underline{A}$, $|\underline{A}|$ *ordered by inclusion is a cpo.*

Proof The bottom element is $\emptyset$. It is closed under entailment because to have $X \vdash Y$, X must be non-empty. Clearly $(|\underline{A}|, \subseteq)$ is a partial order.

Let S be a directed subset of $|\underline{A}|$. It is easy to check that $\bigcup S \in |\underline{A}|$; this is because for any finite subset X of $\bigcup S$ there is an $s \in S$ such that $X \subseteq s$, by the directedness of S and finiteness of X. It is then easy to see that $\bigcup S$ is closed under entailment.

□

Although for directed sets S we have $\bigsqcup S = \bigcup S$, this is not necessarily true when S is not directed. For undirected but bounded S, the least upper bound may not exist. In the definition of a sequent structure $(A, \vdash)$ there is no restriction on the set A. Accordingly $(|\underline{A}|, \subseteq)$ need not be ω-algebraic. However, even restricting A to countable, $|\underline{A}|$ still need not be ω-algebraic since it can be that the minimal upper bounds are compact, but uncountable. Furthermore, $(|\underline{A}|, \subseteq)$ need not be algebraic.

Since our concern is to find a certain category of sequent structures to represent SFP domains, we will not go further into what kind of domains are represented by general sequent structures. That important topic is discussed in [DG90a], where a structure called *a non-deterministic information system* is introduced and a topological characterization is provided for the kind of cpos represented. It turns out that sequent structures and non-deterministic information systems represent the same class of cpos, although the structures look different. This means some of the results in

[DG90a] can be applied here directly. In particular, examples can be found in that paper which back up our claims made at the end of the previous paragraph.

The following example [DG90a] shows that sequent structures can represent non-algebraic domains.

Example 3.1 *Let* $\underline{A} = (A, \vdash)$ *be a sequent structure with*

$$A = \{a_i \mid i \in \omega\} \cup \{b_i \mid i \in \omega\},$$

and $\vdash$ *generated by*

$$\begin{aligned} &a_i \vdash a_j,\ a_i \vdash b_j,\ b_i \vdash b_j \quad \textit{if } j \leq i, \\ &b_i \vdash \{a_j, b_k\} \quad \textit{if } j \leq i,\ k \in \omega, \quad \textit{and} \\ &\{a_j, b_i\} \vdash a_i \quad \textit{for all } i, j \in \omega. \end{aligned}$$

Then the elements of $\underline{A}$ *are:*

$$\begin{aligned} &y = \{b_i \mid i \in \omega\}, \\ &x_i = \{a_j \mid j < i\} \cup \{b_j \mid j < i\}, \quad \textit{for each } i \in \omega, \textit{ and} \\ &A \ \textit{ itself.} \end{aligned}$$

The cpo determined by this sequent structure can be pictured as:

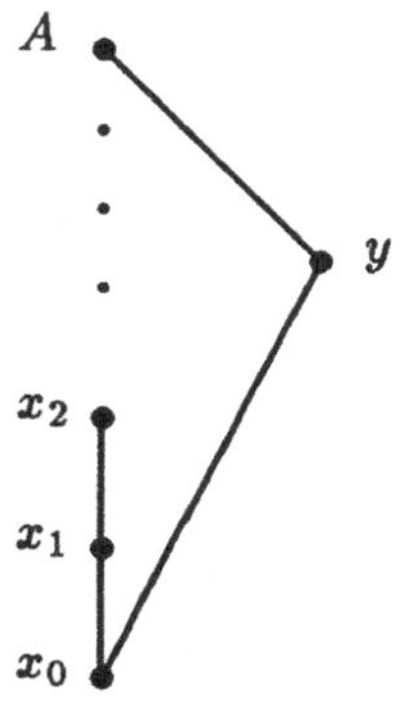

which is not algebraic.

3.2 Information Systems

This section discusses the relationship of sequent structures and information systems. An information system is a structure which consists of a set of propositions, a consistency predicate, and an entailment relation specified as follows:

Definition 3.3 *An information system is a structure* $\underline{A} = (A, Con, \vdash)$ *where*

- A *is a set of propositions,*
- $Con \subseteq \{B \mid B \Subset A\}$ *is the consistent sets,*
- $\vdash \subseteq Con \times A$, *the entailment relation,*

which satisfy

1. $(X \subseteq Y \;\&\; Y \in Con) \Longrightarrow X \in Con$,
2. $a \in A \Longrightarrow \{a\} \in Con$,
3. $(X \vdash a \;\&\; X \in Con) \Longrightarrow X \cup \{a\} \in Con$,
4. $(a \in X \;\&\; X \in Con) \Longrightarrow X \vdash a$,
5. $(\forall b \in Y. X \vdash b \;\&\; Y \vdash c) \Longrightarrow X \vdash c$.

Here $X \nvdash Y$ stands for $\neg(X \vdash Y)$. Clearly an information system $(A, \vdash)$ can be viewed as a sequent structure $(A, \vdash')$ by taking $X \vdash' Y$ if and only if $X \vdash b$ for some b in Y, and $X \vdash' \emptyset$ if and only if $X \notin Con$. On the other hand, *deterministic sequent structures* give information systems.

Definition 3.4 *A sequent structure* $\underline{A}$ *is called deterministic if it satisfies the following:*

- $(X \vdash Y \& X \nvdash \emptyset) \Longrightarrow \exists b \in Y \,.\, X \vdash b$,
- $\forall a \in A.\, a \nvdash \emptyset$.

The first axiom has the effect that every entailment $X \vdash Y$ with $X \nvdash \emptyset$ is a consequence of an entailment of the form $X \vdash b$ with $b \in Y$. The second axiom says that every single proposition is consistent. Such sequent structures determine information systems.

Theorem 3.3 *A deterministic sequent structure determines an information system.*

Proof Let $(A, \vdash)$ be a deterministic sequent structure.

Define *Con* to be the set $\{X \mid X \not\vdash \emptyset\}$. We show that the five axioms for an information system are derivable from the sequent structure. It is sufficient to show that

1. $X \subseteq Y \;\&\; Y \not\vdash \emptyset \Longrightarrow X \not\vdash \emptyset$,
2. $a \not\vdash \emptyset$,
3. $X \vdash a \;\&\; X \not\vdash \emptyset \Longrightarrow X, a \not\vdash \emptyset$,
4. $a \in X \Longrightarrow X \vdash a$,
5. $(\forall b \in Y. X \vdash b \;\&\; Y \vdash c) \Longrightarrow X \vdash c$.

The first statement follows form weakening. 2 and 4 are trivial. 3 follows from the fifth conclusion of Theorem 3.1 and 5 from the second conclusion of Theorem 3.1.

□

Similarly, an information system determines a family of elements. As a special case of Definition 3.2, we have (recall that $X \Subset x$ means X is a finite subset of x):

Definition 3.5 *The elements* $\mid \underline{A} \mid$, *of an information system*

$$\underline{A} = (A, \; Con, \; \vdash)$$

consists of subsets x of propositions which are

- *finitely consistent:* $X \Subset x \Longrightarrow X \in Con$, *and*
- *closed under entailment:* $X \subseteq x \;\&\; X \vdash a \Longrightarrow a \in x$.

Therefore, for a deterministic sequent structure $\underline{A}$, $(\mid \underline{A} \mid, \subseteq)$ is a Scott domain by the results of [Sc82] and [LW84].

3.3 Strongly Finite Sequent Structures

This section introduces strongly finite sequent structures. They are then shown to represent SFP domains. From now on we assume that propositions are countable. Given a sequent structure $(A, \vdash)$, we abbreviate

$$Y \Subset A \;\&\; \forall b \in Y.\, X \vdash b \qquad \text{as} \qquad X \vdash_{\wedge} Y$$

and

$$X \vdash Y \;\&\; \forall b \in Y.\, b \vdash_\wedge X \qquad \text{as} \qquad X \approx Y.$$

One can read $X \vdash_\wedge Y$ as *conjunction of X's entails conjunction of Y's*, and $X \approx Y$ as *conjunction of X's is equivalent to disjunction of Y's*. Note that $\approx$ is *not symmetric*.

Definition 3.6 *A SS $\underline{A}$ is called strongly finite (SFSS) if it satisfies the axiom of finite closure: for all $X \Subset A$ there is a finite super set $Y \supseteq X$ such that*

$$(Y \supseteq Y' \neq \emptyset) \Longrightarrow \exists Y'' \subseteq Y\,.\, Y' \approx Y'',$$

as well as the atomicity axiom: $a \vdash X \Longrightarrow \exists b \in X.\, a \vdash b$.

The atomicity axiom implies that for any proposition a, it can never be the case that $a \vdash \emptyset$. That is to say, every single proposition is consistent. The axiom of finite closure says that for any finite set of propositions there is a super set which has the property that any conjunction of propositions of a subset of the super set is equivalent to a disjunction of propositions of some subset of the super set. This is really a restatement of quasi-conjunctive closedness.

Proposition 3.1 *Let $\underline{A}$ be a SFSS and let $X \vdash Y$. Then there is some $Y' \subseteq Y$ such that $X \vdash Y'$ and*

$$\forall b,\, b' \in Y'.\, (b \vdash b' \Longrightarrow b = b').$$

This proposition says that if $X \vdash Y$, it may be possible to make Y smaller by getting rid of some relatively weaker propositions in Y.

Proof We can get such a Y' by repeated use of the third conclusion of Theorem 3.1: $X \vdash Y'$ and $b \vdash b'$ with $b,\, b' \in Y'$ implies

$$X \vdash (Y' \setminus \{b\}) \cup \{b'\},$$

or $X \vdash Y' \setminus \{b\}$. Thus each such step reduces the number of elements of Y' by one if $b \neq b'$ and $b \vdash b'$, and we must stop somewhere because Y' is finite (the argument for the case where Y is already an empty set is trivial).

□

Suppose $X \approx Z$ and $X \vdash \emptyset$. We have $X \vdash Z$ and $b \vdash_\wedge X$ for all $b \in Z$. Therefore $b \vdash \emptyset$ for all $b \in Z$ because of the second conclusion of Theorem 3.1. This is possible only when Z is empty.

We remark that if a SS $\underline{A}$ is deterministic, it is not necessarily strongly finite. However, if $\underline{A}$ is deterministic and *closed under conjunction* in the sense that $X \not\vdash \emptyset$ implies $X \approx a$ for some $a \in A$, then it is clearly also strongly finite.

Theorem 3.4 *Let $\underline{A}$ be a strongly finite sequent structure. Then*

1. $(|\underline{A}|, \subseteq)$ *is a cpo.*
2. *For each $a \in A, \overline{a} = \{ b \mid b \in A \;\&\; a \vdash b \}$ is a finite element of $|\underline{A}|$.*
3. $(|\underline{A}|, \subseteq)$ *is ω-algebraic.*
4. *For a finite set S of finite elements of $|\underline{A}|$, $\bowtie S$, the set of minimal upper bounds of S, is finite and complete.*

Proof (1): Straightforward. (2): Let $X \Subset \overline{a}$ and $X \vdash Y$. Then $a \vdash_{\wedge} X$. Hence $a \vdash Y$ by the second conclusion of Theorem 3.1. By the atomicity of SFSS, there is some $c \in Y$ such that $a \vdash c$, which implies $c \in \overline{a}$. Therefore $\overline{a}$ is an element. It is finite because for any directed $S \subseteq |\underline{A}|$, if $\overline{a} \subseteq \bigsqcup S$, *i.e.*, $\overline{a} \subseteq \bigcup S$, then we have $a \in \bigcup S$. Hence $a \in s$ for some $s \in S$, and $\overline{a} \subseteq s$. (3): Note that for every $x \in |\underline{A}|$, $x = \bigsqcup\{ \overline{a} \mid a \in x \cap A \}$. By Lemma 2.1, all the non-bottom finite elements are of the form $\overline{a}$. Hence $(|\underline{A}|, \subseteq)$ is ω-algebraic. (4): Let S be a finite, consistent subset of finite elements of $|\underline{A}|$. We know from the proof of (3) that $S = \{\overline{a} \mid a \in X\}$ for some $X \Subset A$ (when $\bot \in S$ it is easy to handle). By the axiom of finite closure there is some finite Y such that $X \approx Y$. Let Y' be a subset of Y such that $\forall b,\ b' \in Y'.\ b \vdash b' \Rightarrow b = b'$. The existence of Y' is assured by Proposition 3.1. It is not difficult to check that $\bowtie S = \{\overline{b} \mid b \in Y'\}$. Hence $\bowtie S$ is finite. It is complete because assuming $x \supseteq s$ for all $s \in S$ we have $x \supseteq X$. Since $X \vdash Y$, there is some b which belongs to $x \cap Y$. Therefore $x \supseteq \overline{b}$.

□

Now we are in a position to show that SFSSs represent SFP domains.

Theorem 3.5 *If $\underline{A}$ is a strongly finite sequent structure then $(|\underline{A}|, \subseteq)$ is an SFP domain.*

Proof Let $\underline{A}$ be a strongly finite sequent structure. It is enough to show that for a finite subset of finite elements S, $U^*(S)$ is finite. We know that S can be written as $S = \{\overline{a} \mid a \in X\}$ for some $X \Subset A$. By the axiom of finite closure there is some finite $R \supseteq X$ such that for any X',

$$(R \supseteq X' \;\&\; X' \neq \emptyset) \Longrightarrow \exists Y \subseteq R \,.\, X' \approx Y.$$

By Theorem 3.4, $\{\overline{a} \mid a \in R\}$ is a $\bowtie$-closed set. Therefore

$$U^*(S) \subseteq \{\overline{a} \mid a \in R\}$$

and $U^*(S)$ is finite.

□

On the other hand, an SFP domain determines a strongly finite sequent structure by taking prime open sets as propositions.

Definition 3.7 *Let D be SFP. Define*

$$IS\,(D) = (\mathbf{P\Omega}(D), \vdash)$$

where $\mathbf{P\Omega}(D)$ is the set $\{\uparrow d \mid d \in D^0 \setminus \{\bot_D\}\}$ and $\vdash$ is a relation on the set of finite subsets of $\mathbf{P\Omega}(D)$ such that

$$X \vdash Y \;\; \text{if} \bigcap X \subseteq \bigcup Y.$$

Note that we do not include D itself as a proposition.

Proposition 3.2 *Let D be an SFP domain. $IS\,(D)$ is an SFSS.*

Proof It is routine to show that $IS\,(D)$ is a SS. When $\bigcap X = \bigcup Y$ we have both $\bigcap X \subseteq \bigcup Y$ and $b \subseteq \bigcup Y \subseteq a$ for any $a \in X$, $b \in Y$, *i.e.*, $X \approx Y$. Hence the axiom of finite closure follows from Theorem 2.6 and Proposition 2.13.

□

Definition 3.8 *Two SFSSs $\underline{A} = (A, \vdash_{\underline{A}})$ and $\underline{B} = (B, \vdash_{\underline{B}})$ are said to be isomorphic if there is a bijection $\beta : (A/{=_{\underline{A}}}) \to (B/{=_{\underline{B}}})$ such that*

$$X \vdash_{\underline{A}} Y \;\; \text{if and only if} \;\; \beta X \vdash_{\underline{B}} \beta Y,$$

where $\beta X = \bigcup_{c \in X}\{a \mid a =_{\underline{A}} \beta(c)\}$, and βY is given similarly; $A/{=_{\underline{A}}}$ is the quotient set and $a =_{\underline{A}} b$ is an abbreviation for $a \vdash_{\underline{A}} b$ and $b \vdash_{\underline{A}} b$.

It is obvious that if $\underline{A}$ and $\underline{B}$ are isomorphic then $\mid \underline{A} \mid$ and $\mid \underline{B} \mid$ are isomorphic.

Proposition 3.3 *For any SFSS $\underline{A}$, $IS\,(\mid \underline{A} \mid)$ is isomorphic to $\underline{A}$, and for any SFP domain D, $\mid IS\,(D) \mid$ is isomorphic to D.*

Proof We give the isomorphism pairs. It is routine to check that they are indeed isomorphisms. The first pair is (θ_1, ϕ_1), where

$\theta_1 : D \rightarrow \mid IS\,(D) \mid$ is given by $e \longmapsto \{ \uparrow d \mid d \sqsubseteq e \;\&\; d \in D^0 \setminus \{ \bot_D \} \}$,
$\phi_1 : \mid IS\,(D) \mid \rightarrow D$ is given by $x \longmapsto \bigsqcup \{ e \mid \uparrow e \in x \}$,

and the second pair is (θ_2, ϕ_2), where

$$\theta_2 : \underline{A} \rightarrow IS\,(\mid \underline{A} \mid) \text{ is given by } a \longmapsto \uparrow \overline{a},$$
$$\phi_2 : IS\,(\mid \underline{A} \mid) \rightarrow \underline{A} \text{ is given by } \uparrow \overline{a} \longmapsto a.$$

□

For information systems in [Sc82] and [LW84], if one starts from $\underline{A}$ and gets back an information system from $\mid \underline{A} \mid$ as described in Definition 3.7, $\underline{A}$ and $IS\,(\mid \underline{A} \mid)$ need not be isomorphic in our sense. Consider, for an example, the information system $(\{0, 1\}, Con, \vdash)$, where Con is th consistency predicate such that $\{0, 1\} \in Con$ and $\vdash$ is the trivial entailment $\{0, 1\} \vdash 0$ and $\{0, 1\} \vdash 1$. The strongly finite sequent structures are canonical, in the sense that the propositions exactly correspond to the non-bottom finite elements of the SFP domain so that $\underline{A}$ and $IS\,(\mid \underline{A} \mid)$ are isomorphic.

3.4 A Category of SFSS

In this section we introduce morphisms on SFSSs called approximable mappings. This makes SFSSs a category. Approximable mappings show how sequent structures relate to one another and they capture the corresponding continuous functions between the associated SFP domains. The way approximable mappings defined is slightly different from the traditional one on information systems. The canonical character of the strongly finite sequent structures makes it possible to specify approximable mappings as relations on propositions rather than on consistent sets.

Definition 3.9 *Let $\underline{A} = (A, \vdash_{\underline{A}})$, $\underline{B} = (B, \vdash_{\underline{B}})$ be SFSSs. An approximable mapping from $\underline{A}$ to $\underline{B}$ is a relation $R \subseteq A \times B$ which satisfies*

$$\forall S \Subset R \quad \forall X \forall Y.$$
$$\pi_1 S \approx_{\underline{A}} X \;\&\; \pi_2 S \approx_{\underline{B}} Y \Longrightarrow \forall a'' \in X \exists b'' \in Y.\, a'' R\, b''.$$

Notice the similarity and difference between this definition and Definition 2.1. For domains the set of minimal upper bounds of any set exists, and is unique. Reflected in Definition 2.1 this fact implies that the quantifications for X and Y there does not really matter. For SFSSs, however, a similar kind of uniqueness does not hold. Therefore universal quantifications are used here for X and Y, to make the approximable mappings 'saturated'.

Note that for an approximable mapping R we have

$$a \vdash a' \;\&\; a' R\, b' \Longrightarrow a R\, b'.$$

This is because when $a \vdash a'$ we have $a' \approx \{a, a'\}$. The desired property now follows from Definition 3.9 by taking $\{(a', b')\} \subseteq R$, $X = \{a, a'\}$, and $Y = \{b'\}$.

Proposition 3.4 *Strongly finite sequent structures with approximable mappings form a category, written* **SFSS**.

Proof Identities are given by $a\, Id_{\underline{A}}\, b$ if $a \vdash_{\underline{A}} b$. We check that approximable mappings compose. Other axioms for a category can be checked similarly.

Let $\underline{A}$, $\underline{B}$ and $\underline{C}$ be SFSSs and $R : \underline{A} \to \underline{B}$ and $S : \underline{B} \to \underline{C}$ be approximable mappings. Let $R \circ S$ be the relational composition. We show that $R \circ S$ is an approximable mapping. Suppose, for a finite set I, $\forall i \in I.\, a_i\,(R \circ S)\, c_i$ and $\{a_i \mid i \in I\} \approx_{\underline{A}} X$, $\{c_i \mid i \in I\} \approx_{\underline{C}} Z$. There exists a $u_i \in B$ such that $a_i\, R\, u_i$, $u_i\, S\, c_i$ for any $i \in I$. Let $\{u_i \mid i \in I\} \approx_{\underline{B}} Y$. The existence of such Y follows from the strong finiteness. Since R is an approximable mapping, $\forall p \in X \exists q \in Y.\, p\, R\, q$. But for each $q \in Y$ we have $q\, S\, c_i$ for all $i \in I$. Therefore there exists some $r \in Z$ such that $q\, S\, r$, since $q \approx q$. Hence $\forall p \in X \exists r \in Z.\, p\,(R \circ S)\, r$.

□

Proposition 3.5 *Let R be an approximable mapping from $\underline{A}$ to $\underline{B}$. Define $f_R : |\underline{A}| \to |\underline{B}|$ by*

$$f_R(x) = \{ b \in B \mid \exists a \in x \,.\, a\,R\,b \}.$$

Then f_R is a continuous function from $|\underline{A}|$ to $|\underline{B}|$.

Proof Let $x \in |\underline{A}|$ and let $R : \underline{A} \to \underline{B}$ be an approximable mapping. To show $f_R(x) \in |\underline{B}|$ let $Y \subseteq f_R(x)$ and $Y \vdash_{\underline{B}} Z$. For each $b \in Y$ there is some $a \in x$ such that $a\,R\,b$. Write X for such a collection of a's. Because $\underline{A}$ and $\underline{B}$ are strongly finite there are X', Y' such that $X \approx_{\underline{A}} X'$ and $Y \approx_{\underline{B}} Y'$. This means we have $X \subseteq x$ and $X \vdash_{\underline{A}} X'$, which implies $X' \cap x \neq \emptyset$. Now let $u_0 \in X' \cap x$. By Definition 3.9 there is a $v_0 \in Y'$ such that $u_0\,R\,v_0$. Thus $v_0 \in f_R(x)$. Also we have $v_0 \vdash_{\underline{B}} Z$, which implies $v_0 \vdash_{\underline{B}} c$ for some $c \in Z$. Therefore $c \in f_R(x) \cap Z$.

The monotonicity of f_R is obvious. It also preserves least upper bounds of directed sets of $|\underline{A}|$; for assuming $b \in f_R(\bigsqcup P)$ with P directed, there is some $a \in \bigcup P$ such that $a\,R\,b$. There is, therefore, some $y \in P$ for which $a \in y$. Hence $b \in f_R(y)$ and

$$f_R(\bigcup P) \subseteq \bigcup \{ f_R(y) \mid y \in P \},$$

enough for the equality to hold.

□

It can be shown that, on the other hand, each continuous function is represented by an approximable mapping. We are not going to prove this directly, however. Instead we show a more general result, that is, that the category of strongly finite sequent structures and the category of SFP domains are equivalent. Informally this result implies that strongly finite information systems and SFP domains are essentially the same. We will use one of MacLane's results ([Ma71], pp 91) in the proof. By this result, a functor F determines an equivalence of the categories if it is full and faithful, and each SFP domain D is isomorphic to $F(\underline{A})$ for some SFSS $\underline{A}$.

Theorem 3.6 **SFSS** *is equivalent to* **SFP**.

Proof Let $F : \mathbf{SFSS} \to \mathbf{SFP}$ be the functor given by

$$F(\underline{A}) = |\underline{A}|$$
$$F(R) = f_R.$$

That each SFP domain D is isomorphic to $F(\underline{A})$ for some SFSS $\underline{A}$ is easy to check. It remains to show that F is full and faithful. First we show that F is full. Let $\underline{A}$ and $\underline{B}$ be SFSSs and $f : F(\underline{A}) \to F(\underline{B})$ a continuous function. Define a relation $R \subseteq A \times B$ by letting $a\,R\,b$ if $b \in f(\overline{a})$. We check that this relation is an approximable mapping form $\underline{A}$ to $\underline{B}$. Let $\{ (a_i, b_i) \mid i \in I \}$ be a finite subset of R. Assume

$$\{ a_i \mid i \in I \} \approx_{\underline{A}} X$$

and

$$\{ b_i \mid i \in I \} \approx_{\underline{B}} Y.$$

For any $a \in X$, $a \vdash_{\wedge} \{ a_i \mid i \in I \}$. Thus we have $b_i \in f(\overline{a_i}) \subseteq f(\overline{a})$ for any $i \in I$. Now $\{ b_i \mid i \in I \} \vdash_{\underline{B}} Y$. Therefore $f(\overline{a}) \cap Y \neq \emptyset$. This means for some $b \in Y$, $b \in f(\overline{a})$, or $a\,R\,b$.

We now show that the continuous function f_R determined by the above R is actually equal to f. Let $x \in \mid \underline{A} \mid$. Suppose $b \in f_R(x)$. By definition there is some $a \in x$, $a\,R\,b$. That is, $b \in f(\overline{a})$. Therefore $b \in f(x)$, by the monotonicity of f. Thus $f_R(x) \subseteq f(x)$. On the other hand, let $b \in f(x)$. By the continuity of f there is some $a \in x$ such that $b \in f(\overline{a})$. Hence $a\,R\,b$ and $b \in f_R(x)$. This means $f(x) \subseteq f_R(x)$. Hence $f = f_R$.

Secondly, we show that F is faithful. Suppose $R, S : \underline{A} \to \underline{B}$ are approximable mappings such that $f_R = f_S$. Let $a\,R\,b$. Then $b \in f_S(\overline{a})$. This means for some $a' \in \overline{a}$, $a'\,S\,b$. By the remark given just before Proposition 3.4, $a\,S\,b$. Therefore, $R \subseteq S$. By symmetry, $S \subseteq R$ and hence $R = S$.

□

Now we consider constructions on SFSSs. We introduce lifting $(\ )_{\uparrow}$, sum $+$, product $\times$ and function space $\to$, with an emphasis on $\to$. The powerdomain constructions are given in the next section. Lifting, sum and product are more or less the same as those on information systems [LW84].

Definition 3.10 *(Lifting) Let $\underline{A} = (A, \vdash)$ be an SFSS. Define the lift of $\underline{A}$ to be $\underline{A}_{\uparrow} = (A', \vdash')$ where*

- $A' = (\{0\} \times A) \cup \{0\}$,
- $X \vdash' Y \Leftrightarrow [0 \in Y \ \ or\ \ \{c \mid (0, c) \in X\} \vdash_{\underline{A}} \{b \mid (0, b) \in Y\}]$.

The lift of an SFSS is formed by joining a new proposition weaker than all the old ones and keeping everything else unchanged.

Definition 3.11 *(Sum) Let $\underline{A} = (A, \vdash_{\underline{A}})$ and $\underline{B} = (B, \vdash_{\underline{B}})$ be SFSSs. Define their sum, $\underline{A} + \underline{B}$, to be $\underline{C} = (C, \vdash)$ where*

- $C = \{0\} \times A \cup \{1\} \times B$,
- $X \vdash Y \Leftrightarrow X = \{(0, a) \mid a \in X_0\} \mathbin{\&} X_0 \vdash_{\underline{A}} \{r \mid (0, r) \in Y\}$ *or*
  $X = \{(1, b) \mid b \in X_1\} \mathbin{\&} X_1 \vdash_{\underline{B}} \{s \mid (1, s) \in Y\}$ *or*
  $X \cap (\{0\} \times A) \neq \emptyset \mathbin{\&} X \cap (\{1\} \times B) \neq \emptyset$.

The sum of two SFSSs is formed by taking a disjoint copy of the propositions, keeping the original entailments on both components, respectively, and making mixed proposition sets inconsistent (*i.e.* entailing everything).

Definition 3.12 *(Product) Let $\underline{A} = (A, \vdash_{\underline{A}})$ and $\underline{B} = (B, \vdash_{\underline{B}})$ be SFSSs. Define their product, $\underline{A} \times \underline{B}$, to be $\underline{C} = (C, \vdash)$ where*

- $C = \{(a, *) \mid a \in A\} \cup \{(*, b) \mid b \in B\} \cup \{(a, b) \mid a \in A \mathbin{\&} b \in B\}$,
- $X \vdash Y \Leftrightarrow \exists c \in Y.\ (\pi_0 X = \{*\} \Rightarrow \pi_0 c = *) \mathbin{\&}$
  $(\pi_1 X = \{*\} \Rightarrow \pi_1 c = *) \mathbin{\&}$
  $\pi_0 X \neq \{*\} \Rightarrow (\pi_0 c = *$ *or* $[\pi_0 X \setminus \{*\}] \vdash_{\underline{A}} \pi_0 c) \mathbin{\&}$
  $\pi_1 X \neq \{*\} \Rightarrow (\pi_1 c = *$ *or* $[\pi_1 X \setminus \{*\}] \vdash_{\underline{B}} \pi_1 c)$.

The symbol $*$ here acts like a proposition which is always true. Note that to have enough propositions for the product the disjoint union of A and B, or the set of pairs $A \times B$, are not sufficient; The use of disjoint union makes $\{(0, a), (1, b)\}$ a non-$\approx$-closed set while the use of $A \times B$ misses out those points corresponding to $(\emptyset, y)$ and $(x, \emptyset)$, which are in $|\underline{A}| \times |\underline{B}|$. The fact that we get enough propositions by introducing $*$ is shown in Theorem 3.7.

Definition 3.13 *For a SFSS $\underline{A} = (A, \vdash)$, $X \Subset A$ is said to be $\approx_{\underline{A}}$ −closed if*

$$(X' \subseteq X \mathbin{\&} X' \neq \emptyset) \Longrightarrow \exists X'' \subseteq X\,.\, X' \approx_{\underline{A}} X''.$$

Thus the axiom of finite closure says every finite set of a SFSS has a finite super set which is $\approx_{\underline{A}}$-closed.

Theorem 3.7 *If $\underline{A}$, $\underline{B}$ are SFSSs then so are $\underline{A}_\uparrow$, $\underline{A}+\underline{B}$, and $\underline{A}\times\underline{B}$. Furthermore,*

$$\begin{aligned} x \in \mid \underline{A}_\uparrow \mid &\Longleftrightarrow (x=\emptyset \ \text{ or } \ \exists y \in \mid \underline{A} \mid .\, x=\{0\}\cup(\{0\}\times y)\,), \\ x \in \mid \underline{A}+\underline{B} \mid &\Longleftrightarrow (\,\exists x_0 \in \mid \underline{A} \mid .\, x_0=\{a \mid (0,a)\in x\}\,) \ \text{ or} \\ &\qquad (\,\exists x_1 \in \mid \underline{B} \mid .\, x_1=\{b \mid (1,b)\in x\}\,), \\ x \in \mid \underline{A}\times\underline{B} \mid &\Longleftrightarrow (a,b)\in x \Rightarrow (a,*),\ (*,b)\in x \ \& \\ &\qquad \{a\in A \mid \exists b\in B.(a,b)\in x\} \in \mid \underline{A} \mid \ \& \\ &\qquad \{b\in B \mid \exists a\in A.(a,b)\in x\} \in \mid \underline{B} \mid . \end{aligned}$$

Proof It is routine to show that $\underline{A}_\uparrow$, $\underline{A}+\underline{B}$, and $\underline{A}\times\underline{B}$ are SSs. To verify the finite closure axiom we need to produce $\approx$-closed sets of propositions. The rules given below indicate how to get them.

$$\begin{aligned} &P \ \approx_{\underline{A}} -\text{closed} \ \& \ \{a \mid (0,a)\in W\} \subseteq P \\ &\quad \Longrightarrow W \subseteq \Sigma_0, \end{aligned}$$

where $\Sigma_0 = \{(0,a) \mid a\in P\} \ \& \ \{(0,a) \mid a\in P\}$ and it is $\approx_{\underline{A}_\uparrow}$-closed;

$$\begin{aligned} &P,Q \ \approx -\text{closed} \ \& \ \{a \mid (0,a)\in W\} \subseteq P \ \& \ \{a \mid (1,a)\in W\} \subseteq Q \\ &\quad \Longrightarrow W \subseteq \Sigma_1, \end{aligned}$$

where $\Sigma_1 = \{(0,a) \mid a\in P\} \cup \{(1,a) \mid a\in Q\}$, which is $\approx_{\underline{A}+\underline{B}}$-closed;

$$\left.\begin{array}{l} P,Q \ \approx -\text{closed} \\ \{a\in A \mid (a,*) \text{ or } (a,b)\in W\} \subseteq P \\ \{b\in B \mid (*,b) \text{ or } (a,b)\in W\} \subseteq Q \end{array}\right\} \Longrightarrow W \subseteq \Sigma_2,$$

where

$$\Sigma_2 = \{(a,*) \mid a\in P\} \cup \{(*,b) \mid b\in Q\} \cup \{(a,b) \mid a\in P \ \& \ b\in Q\}$$

a $\approx_{\underline{A}\times\underline{B}}$-closed set.

The first two rules are obviously valid. To check the third rule let $W' \subseteq \Sigma_2$ We show that there is some $W'' \subseteq \Sigma_2$ such that $W' \approx_{\underline{A}\times\underline{B}} W''$. This is trivial when there are no elements of the form $(a,*)$ or (a,b) in W', and similarly for the case when there are no elements of the form $(*,b)$ or (a,b) in W'. So suppose

$$\{a\in A \mid (a,*)\in W' \text{ or } \exists b\in B.(a,b)\in W'\} \neq \emptyset$$

and

$$\{ b \in B \mid (*, b) \in W' \text{ or } \exists a \in A. (a, b) \in W' \} \neq \emptyset.$$

Since P and Q are $\approx$-closed, there are $P' \subseteq P$, $Q' \subseteq Q$ such that

$$\{ a \in A \mid (a, *) \in W' \text{ or } \exists b \in B. (a, b) \in W' \} \approx_{\underline{A}} P'$$

and

$$\{ b \in B \mid (*, b) \in W' \text{ or } \exists a \in A. (a, b) \in W' \} \approx_{\underline{B}} Q'.$$

Let $W'' = \{ (a, b) \mid a \in P' \ \& \ b \in Q' \}$. Clearly $W' \approx_{\underline{A} \times \underline{B}} W''$.

It is routine to show the second part of the theorem.

□

From this theorem we can see that there is an isomorphism between $| \underline{A} + \underline{B} |$ and $| \underline{A} | + | \underline{B} |$, and $| \underline{A} \times \underline{B} |$ and $| \underline{A} | \times | \underline{B} |$. Thus it justifies our definition of product and sum.

The construction of function space is more complicated. We need an auxiliary definition first.

Definition 3.14 *Let $\underline{A} = (A, \vdash_{\underline{A}})$ and $\underline{B} = (B, \vdash_{\underline{B}})$ be SFSSs. A finite subset m of $A \times B$ is said to be a molecule if*

$$\begin{aligned} \forall m' \subseteq m \exists X \subseteq \pi_1 m \exists Y. \quad & \pi_1 m' \vdash_{\underline{A}} X \\ & \pi_2 m' \vdash_{\underline{B}} Y, \quad \textit{and} \\ & \forall a \in X \exists b \in Y. (a, b) \in m. \end{aligned}$$

Molecules are given as one of the descriptions equivalent to Definition 2.1. They correspond to step functions.

Definition 3.15 *(Function space) Let $\underline{A} = (A, \vdash_{\underline{A}})$ and $\underline{B} = (B, \vdash_{\underline{B}})$ be SFSSs. Their function space, $\underline{A} \to \underline{B}$, is the sequent structure $\underline{C} = (C, \vdash,)$ given by*

- $C = \{ m \mid m \ \textit{is a molecule in } A \times B \}$,
- $X \vdash Y$ *iff* $(\forall m' \in X. \{ m \} \vdash \{ m' \}) \Longrightarrow \exists m'' \in Y. \{ m \} \vdash \{ m'' \}$

  where $\{ m \} \vdash \{ m' \} \Longleftrightarrow$

  $\forall \alpha' \in m' \exists \alpha \in m. \{ \pi_1 \alpha' \} \vdash_{\underline{A}} \{ \pi_1 \alpha \} \ \& \ \{ \pi_2 \alpha \} \vdash_{\underline{B}} \{ \pi_2 \alpha' \}$.

There is a general guideline according to which we can test the definition of entailment. $X \vdash Y$ in a sequent structure should mean

$$\bigcap\{\uparrow\overline{a} \mid a \in X\} \subseteq \bigcup\{\uparrow\overline{b} \mid b \in Y\}$$

in the topology of the domain. In other words, for any point x above all $\overline{a}$ with $a \in X$, this x should be above some point $\overline{b}$ for $b \in Y$. Since the domains concerned are algebraic, this x can be chosen from the set of finite elements, which are of the form $\overline{a}$.

It is reasonable to have, for molecules m, m',

$$\begin{aligned}&\{m\} \vdash \{m'\} \iff \\ &\forall \alpha' \in m' \, \exists \alpha \in m. \, \{\pi_1\alpha'\} \vdash_{\underline{A}} \{\pi_1\alpha\} \; \& \; \{\pi_2\alpha\} \vdash_{\underline{B}} \{\pi_2\alpha'\}.\end{aligned}$$

Therefore,

$$(\forall m' \in X. \{m\} \vdash \{m'\}) \Longrightarrow \exists m'' \in Y. \{m\} \vdash \{m''\}$$

is just a direct translation of the corresponding topological relation.

The following proposition shows that the construction of function space gives a strongly finite sequent structure.

Proposition 3.6 *If $\underline{A}$ and $\underline{B}$ are strongly finite then $\underline{A} \to \underline{B}$ is strongly finite.*

Proof First $\underline{A} \to \underline{B}$ is a SS. This is a routine task of checking the corresponding rules.

To show that $\underline{A} \to \underline{B}$ is strongly finite one can make use of the fact that molecules corresponding to step functions in the function space, the entailment on molecules reflects the order in the function space, and the function space itself has a $\bowtie$-closed super set for every finite set of step functions.

□

The following isomorphism is expected.

Proposition 3.7 *If $\underline{A}$ and $\underline{B}$ are SFSSs then*

$$|\underline{A} \to \underline{B}| \cong |\underline{A}| \to |\underline{B}|.$$

Proof Straightforward.

□

3.5 Powerdomains

This section introduces the Hoare, the Smyth, and the Plotkin powerdomain constructions on SFSSs.

Definition 3.16 *Let $\underline{A}$ be a strongly finite sequent structure. The Hoare powerdomain of $\underline{A}$ is the sequent structure $P_H\underline{A} = (C, \vdash)$ where*

- $C = \{B \mid B \Subset A \ \&\ B \neq \emptyset\}$,
- $X \vdash Y \iff \exists \beta \in Y. (\forall b \in \beta \exists a \in \bigcup X.\ a \vdash_{\underline{A}} b)$.

According to the definition, $\{\alpha\} \vdash \{\beta\}$ if and only if

$$\forall b \in \beta \exists a \in \alpha.\ a \vdash_{\underline{A}} b.$$

Therefore $\vdash$ on C is a reverse of the preorder $\sqsubseteq_1$ introduced in Section 2.6. We want $X \approx Y$ to characterize the situation

$$\bigcap\{\uparrow\overline{\alpha} \mid \alpha \in X\} \subseteq \bigcup\{\uparrow\overline{\beta} \mid \beta \in Y\}.$$

This is equivalent to saying that whenever $\{\alpha_0\} \vdash \{\alpha\}$ for all $\alpha \in X$, $\{\alpha_0\} \vdash \{\beta_0\}$ for some $\beta_0 \in Y$. Clearly the entailment thus derived is equivalent to $X \vdash Y$ since we have $X \approx \{\bigcup X\}$.

We can think of $\alpha \in C$ as a logical formula $\bigwedge\{\Diamond a \mid a \in \alpha\}$, where intuitively a set S of processes satisfies $\Diamond a$ if there exists $p \in S$, p satisfies a. Then we have, for example,

$$\left(\bigwedge\{\Diamond a \mid a \in \alpha\}\right) \wedge \left(\bigwedge\{\Diamond b \mid b \in \beta\}\right) \iff \bigwedge\{\Diamond c \mid c \in \alpha \cup \beta\},$$

in the sense that a set of processes satisfies the proposition on the left hand side if and only if it satisfies the proposition on the right hand side.

Proposition 3.8 *If $\underline{A}$ is a strongly finite sequent structure then so is $P_H\underline{A}$.*

Proof $P_H\underline{A}$ is clearly a SS. Suppose $X \Subset C$. Let $\beta = \bigcup X$. Then $\beta \in C$, $X \vdash \{\beta\}$, and $\forall \alpha \in X\ \{\beta\} \vdash \{\alpha\}$ according to the definition of entailment. Therefore $X \approx \{\beta\}$, from which it is easy to see that the finite closure axiom holds.

□

The definition of entailment for $P_H\underline{A}$ implies that, as explained earlier, if $X \vdash Y$ then there is some $\beta \in Y$ such that $X \vdash \{\beta\}$. Thus $P_H\underline{A}$ is actually deterministic. This indicates that $\mid P_H\underline{A} \mid$ is always a Scott domain.

Proposition 3.9 *$|P_H\underline{A}|$ is isomorphic to $\mathcal{P}_H|\underline{A}|$.*

Proof We want to establish an order preserving one-one correspondence between the finite elements of $|P_H\underline{A}|$ and $\mathcal{P}_H|\underline{A}|$. Define

$$\theta : |P_H\underline{A}| \to \mathcal{P}_H|\underline{A}| \qquad \overline{\alpha} \longmapsto Cl_H(\{\overline{a} \mid a \in \alpha\})$$

with $\alpha \in C$ and

$$\eta : \mathcal{P}_H|\underline{A}| \to |P_H\underline{A}| \qquad Cl_H(\{\overline{a} \mid a \in X\}) \longmapsto \overline{X},$$

where $X \Subset A$. Suppose $\overline{\alpha} \subseteq \overline{\beta}$. Then $\alpha \in \overline{\beta}$, or $\{\beta\} \vdash \{\alpha\}$. But by definition we have $\forall a \in \alpha \exists b \in \beta. \{b\} \vdash_{\underline{A}} \{a\}$. Hence

$$\forall \overline{a} \in \{\overline{a'} \mid a' \in \alpha\} \exists \overline{b} \in \{\overline{b'} \mid b' \in \beta\}. \overline{a} \subseteq \overline{b}.$$

Therefore $\{\overline{a'} \mid a \in \alpha\} \sqsubseteq_1 \{\overline{b'} \mid b' \in \beta\}$, which implies that θ is order preserving. It is easy to see that θ is one-one. That $\eta \circ \theta = \mathrm{id}_{|P_H\underline{A}|}$ and $\theta \circ \eta = \mathrm{id}_{\mathcal{P}_H|\underline{A}|}$ are also obvious.

□

We now deal with the Smyth powerdomain.

Definition 3.17 *Let $\underline{A}$ be a strongly finite sequent structure. The Smyth powerdomain of $\underline{A}$ is the sequent structure $P_S\underline{A} = (C, \vdash)$ where*[1]

- $C = \{B \mid B \Subset A \ \&\ B \neq \emptyset\}$,
- $X \vdash Y \iff \exists \beta \in Y. \forall \alpha_0 \in C. (\forall \alpha \in X. \alpha_0 \cap \alpha \neq \emptyset) \Longrightarrow \alpha_0 \vdash_{\underline{A}} \beta$.

When $X = \{\alpha\}, Y = \{\beta\}$ where $\alpha, \beta \in C$, the entailment $\{\alpha\} \vdash \{\beta\}$ means $\forall W \Subset A. (W \cap \alpha \neq \emptyset \Longrightarrow \forall w \in W \exists b \in \beta. \{w\} \vdash_{\underline{A}} \{b\})$. In other words, $\{\alpha\} \vdash \{\beta\}$ if and only if $\forall a \in \alpha \exists b \in \beta. \{a\} \vdash \{b\}$. Therefore $\vdash_{P_S\underline{A}}$ on C is a reverse of the preorder $\sqsubseteq_0$ introduced in Chapter 2 on the non-empty, finite sets of finite elements of $|\underline{A}|$.

Having agreed on what the entailment should be on singletons, we check that $X \vdash Y$ is equivalent to

$$\forall \alpha_0 \in C. (\forall \alpha \in X. \{\alpha_0\} \vdash \{\alpha\}) \Longrightarrow \exists \beta \in Y. \{\alpha_0\} \vdash \{\beta\},$$

[1] In an extended version of [Sc82], Scott call a finite set α_0 with the property $(\forall \alpha \in X. \alpha_0 \cap \alpha \neq \emptyset)$ a *choice set*.

which we write as $X \vdash' Y$, a description of the situation

$$\bigcap\{\uparrow\overline{\alpha} \mid \alpha \in X\} \subseteq \bigcup\{\uparrow\overline{\beta} \mid \beta \in Y\}.$$

$X \vdash Y \Longrightarrow X \vdash' Y$: Suppose $\alpha_0 \in C$ and $\forall \alpha \in X. \{\alpha_0\} \vdash \{\alpha\}$. Let $\beta_0 \in Y$ be such that

$$(\forall \alpha \in X. \alpha' \cap \alpha \neq \emptyset) \Longrightarrow \alpha' \vdash_{\underline{A}} \beta_0.$$

Let $a_0 \in \alpha_0$. For any $\alpha \in X$, there is $b_\alpha \in \alpha$ for which $\{a_0\} \vdash_{\underline{A}} \{b_\alpha\}$. Clearly $\{b_\alpha \mid \alpha \in X\} \cap \alpha \neq \emptyset$ for all $\alpha \in X$. Therefore,

$$\{b_\alpha \mid \alpha \in X\} \vdash_{\underline{A}} \beta_0.$$

However, $\{a_0\} \vdash_\wedge \{b_\alpha \mid \alpha \in X\}$. Thus $\{a_0\} \vdash_{\underline{A}} \beta_0$ which implies $\{a_0\} \vdash_{\underline{A}} \{b_0\}$ for some $b_0 \in \beta_0$ and this is true for any a_0 in α_0, which means $\{\alpha_0\} \vdash \{\beta_0\}$.

$X \vdash' Y \Longrightarrow X \vdash Y$: Suppose $X \vdash' Y$. Rewrite X as $\{\alpha_i \mid i \in I\}$, where I is finite. For any

$$f \in \{g : I \to \bigcup X \mid \forall i \in I. g(i) \in \alpha_i\},$$

let $\{f(i) \mid i \in I\} \approx_{\underline{A}} Z_f$. Clearly for any c in $\bigcup_f Z_f$, for any $i \in I$, $\{c\} \vdash_{\underline{A}} \alpha_i$. Therefore for any i, $\{\bigcup_f Z_f\} \vdash \{\alpha_i\}$, and hence $\{\bigcup_f Z_f\} \vdash \{\beta_0\}$ for some $\beta_0 \in Y$, since $X \vdash' Y$. Now let $\alpha_0 \in C$ be such that for any α in X, $\alpha_0 \cap \alpha \neq \emptyset$. For each i, select an $a_i \in \alpha_0 \cap \alpha_i$. The collection of such a_i's corresponds to a function h such that $h(i) = a_i$. If $\alpha_0 \vdash_{\underline{A}} \emptyset$ then there is nothing to prove; otherwise we have $\{h(i) \mid i \in I\} \approx_{\underline{A}} Z_h$ and $Z_h \neq \emptyset$. $\{\bigcup_f Z_f\} \vdash \{\beta_0\}$ implies $\{c\} \vdash_{\underline{A}} \beta_0$ for all $c \in Z_h$. Hence $\alpha_0 \vdash_{\underline{A}} \beta_0$.

From the above explanation it can be seen that we could have used the following definition of entailment for the Smyth powerdomain:

$$X \vdash Y \text{ iff } \exists \beta \in Y. \forall \alpha_0 \subseteq \bigcup X. (\forall \alpha \in X. \mid \alpha_0 \cap \alpha \mid = 1) \Longrightarrow \alpha_0 \vdash_{\underline{A}} \beta.$$

It is slightly better, since it avoids the use of universal quantifier over C.

It is suitable to think of $\alpha \in C$ as $\Box \bigvee \alpha$, with the interpretation that a set of processes S satisfies $\Box \bigvee \alpha$ if each process in S satisfies $\bigvee \alpha$. We have, under this interpretation, $\Box \bigvee X \Longrightarrow \Box \bigvee Y$ if and only if $\bigvee X \Longrightarrow \bigvee Y$, if and only if $\forall a \in X\ \exists b \in Y.\ a \Rightarrow b$ and

$$\begin{aligned}&(\Box \bigvee X_1) \wedge (\Box \bigvee X_2) \wedge \cdots (\Box \bigvee X_n)\\ \Longleftrightarrow\ &\Box[(\bigvee X_1) \wedge (\bigvee X_2) \wedge \cdots (\bigvee X_n)].\end{aligned}$$

Proposition 3.10 *If $\underline{A}$ is a strongly finite sequent structure then so is $P_S\underline{A}$.*

Proof The only non-trivial part is finite closure. Let V be a finite set of propositions of $P_S\underline{A}$. $\bigcup V$ is a finite set of propositions of $\underline{A}$. Therefore there is a finite set P of $\underline{A}$, $\approx_{\underline{A}}$-closed and contains $\bigcup V$. Then $\{\alpha \mid \alpha \subseteq P\}$ is a $\approx$-closed set of $P_S\underline{A}$. In fact, let $X \subseteq \{\alpha \mid \alpha \subseteq P\}$. Rewrite X as $\{\alpha_i \mid i \in I\}$, where I is finite. For any

$$f \in \{g : I \to \bigcup X \mid \forall i \in I.\, g(i) \in \alpha_i\},$$

let $\{f(i) \mid i \in I\} \approx_{\underline{A}} Z_f$. Clearly for any c in $\bigcup_f Z_f$, for any $i \in I$, $\{c\} \vdash_{\underline{A}} \alpha_i$. Therefore for any i, $\{\bigcup_f Z_f\} \vdash \{\alpha_i\}$. According to our definition of entailment for $P_S\underline{A}$, $X \approx \{\bigcup_f Z_f\}$.

□

From the proof we can see that, $P_S\underline{A}$, too, is deterministic. Similarly we have the following proposition, whose proof is omitted.

Proposition 3.11 *$|P_S\underline{A}|$ is isomorphic to $\mathcal{P}_S|\underline{A}|$.*

We take a similar approach to the Plotkin powerdomain. However, this construction is more complicated, especially the related proofs. Perhaps one of the reasons for this is that it is this construction that does not preserve consistent completeness of Scott domains.

Definition 3.18 *Let $\underline{A}$ be an SFSS. The Plotkin powerdomain of $\underline{A}$ is the sequent structure $P_P\underline{A} = (C, \vdash)$, with*

- $C = \{B \mid B \Subset A \;\&\; B \neq \emptyset\}$,
- $X \vdash Y$ *iff* $\forall \beta. (\forall \alpha \in X. \{\beta\} \vdash \{\alpha\}) \Longrightarrow \exists \beta' \in Y. \{\beta\} \vdash \{\beta'\}$

  where $\{\alpha\} \vdash \{\beta\}$ *iff* $\{\alpha\} \vdash_{P_H\underline{A}} \{\beta\} \;\&\; \{\alpha\} \vdash_{P_S\underline{A}} \{\beta\}$.

The following is expected.

Proposition 3.12 *If $\underline{A}$ is an SFSS then so is $P_P\underline{A}$.*

Proof $P_P\underline{A}$ is clearly a SS. Let $P_P\underline{A} = (C, \vdash)$. Then

$$\begin{aligned} V \Subset C &\Longrightarrow \bigcup V \Subset A \\ &\Longrightarrow \exists P \Subset A.\, \bigcup V \subseteq P \;\&\; P \text{ is } \approx_{\underline{A}}\text{-closed} \\ &\Longrightarrow \{\alpha \mid \alpha \subseteq P\} \text{ is } \approx_{P_P\underline{A}}\text{-closed.} \end{aligned}$$

The last step, that $\{\alpha \mid \alpha \subseteq P\}$ is $\approx_{P_P\underline{A}}$-closed, is the most difficult to check directly. It amounts to showing that for any $X \subseteq \{\alpha \mid \alpha \subseteq P\}$, there is a subset $X' \subseteq \{\alpha \mid \alpha \subseteq P\}$ such that $X \approx_{P_P\underline{A}} X'$. One can design a procedure to derive such an X' from X (such as the one implied by Lemma 4.1 to Lemma 4.5 later). However, the existence of X' is implied by Theorem 3.6 and Theorem 2.9.

□

Furthermore, we have

Proposition 3.13 *$\mid P_P\underline{A} \mid$ is isomorphic to $\mathcal{P}_P \mid \underline{A} \mid$.*

3.6 A Cpo of SFSS

Following the idea described in [LW84] we introduce a complete partial order of SFSSs based on which recursively defined structures can be given. We show that all the constructions induce continuous functions on the cpo. The order on SFSSs is based on an intuitive notion of *substructure*.

Definition 3.19 *Let $\underline{A} = (A, \vdash_{\underline{A}})$, $\underline{B} = (B, \vdash_{\underline{B}})$ be SFSSs. $\underline{A} \trianglelefteq \underline{B}$ if*

- $A \subseteq B$,
- $X \vdash_{\underline{A}} Y \iff X \cup Y \subseteq A \;\&\; X \vdash_{\underline{B}} Y$.

When $\underline{A} \trianglelefteq \underline{B}$ we call $\underline{A}$ a substructure of $\underline{B}$.

Proposition 3.14 *Let $\underline{A}$ and $\underline{B}$ be SFSSs. If $\underline{A} \trianglelefteq \underline{B}$ then there is an embedding-projection pair between $\mid \underline{A} \mid$ and $\mid \underline{B} \mid$.*

Proof Define

$$\theta : \mid \underline{A} \mid \to \mid \underline{B} \mid \qquad x \longmapsto \{ b \mid \exists a \in x . a \vdash_{\underline{B}} b \}$$

and

$$\phi : \mid \underline{B} \mid \to \mid \underline{A} \mid \qquad y \longmapsto y \cap A.$$

Let $x \in \mid \underline{A} \mid$. We check $\theta(x) \in \mid \underline{B} \mid$. Suppose $Z \subseteq \theta(x)$ and $Z \vdash_{\underline{B}} H$. By definition $\forall c \in Z \exists a \in x.\ a \vdash_{\underline{B}} c$. Let X be the collection of such a's. Clearly $X \subseteq x$. $X \vdash_{\wedge} Z$, which implies $X \vdash_{\underline{B}} H$. By the finite closure axiom there is some $X' \subseteq A$ such that $X \approx_{\underline{A}} X'$. As $X \subseteq x$ and $X \vdash_{\underline{A}} X'$, there is some $a' \in x \cap X'$. But $\underline{A}$ is a substructure of $\underline{B}$. We have $X \approx_{\underline{B}} X'$, and hence $a' \vdash a$ for each $a \in X$. Thus $a' \vdash_{\underline{B}} H$ and therefore $a' \vdash_{\underline{B}} b$ for some $b \in H$. That is, $b \in \theta(x)$.

Let $y \in |\underline{B}|$. We check $y \cap A \in |\underline{A}|$. Suppose $Y \subseteq y \cap A$ and $Y \vdash_{\underline{A}} Z$. Then $\exists c \in Z.\, c \in y$. But $Z \subseteq A$, so $c \in y \cap A$.

The proof that θ, ϕ form an embedding-projection pair, *i.e.*, $\phi \circ \theta = \mathrm{id}_{|\underline{A}|}$ and $\theta \circ \phi \sqsubseteq \mathrm{id}_{|\underline{B}|}$, is then straightforward.

□

Similar to what has been pointed out in [LW84], the collection of SFSSs do not form a set but rather a class. Therefore they cannot form a complete partial order(cpo) in the ordinary sense. We could say that they form a *large* cpo $\mathbf{CPO}_{\mathrm{SFSS}}$. Nevertheless the standard theory of fixed-points of continuous functions still works for $\mathbf{CPO}_{\mathrm{SFSS}}$, and that is all we need.

Theorem 3.8 *The relation* $\trianglelefteq$ *on* $\mathbf{CPO}_{\mathrm{SFSS}}$ *is a partial order with the least element* $\bot = (\emptyset, \emptyset)$. *If* $\underline{A}_0 \trianglelefteq \underline{A}_1 \trianglelefteq \cdots \trianglelefteq \underline{A}_i \trianglelefteq \cdots$ *is an increasing chain of SFSSs where* $\underline{A}_i = (A_i, \vdash_i)$, *then their least upper bound is*

$$\bigcup_{i\in\omega} \underline{A}_i = \left(\bigcup_{i\in\omega} A_i,\ \bigcup_{i\in\omega} \vdash_i\right).$$

Proof It is routine to check that

$$\bigcup_{i\in\omega} \underline{A}_i = \left(\bigcup_{i\in\omega} A_i,\ \bigcup_{i\in\omega} \vdash_i\right)$$

is a SFSS. For each i, $\underline{A}_i \trianglelefteq \bigcup_{k\in\omega} \underline{A}_k$ because of the following:

1. $A_i \subseteq \bigcup_{k\in\omega} A_k$.

2. If $X \cup Y \subseteq A_i$ and $X \vdash_{\bigcup_{k\in\omega}\underline{A}_k} Y$ then $X \vdash_j Y$ for some $j \geq i$ because $\vdash_{\bigcup_{k\in\omega}\underline{A}_k} = \bigcup_{k\in\omega} \vdash_k$. Therefore $X \vdash_i Y$.

It is also the least upper bound of the chain. Suppose $\underline{B}$ is an upper bound of the chain. Then $\bigcup_{i\in\omega} \underline{A}_i \trianglelefteq \underline{B}$ since $\bigcup_{i\in\omega} A_i \subseteq B$ and

$$\begin{aligned} X \vdash_{\bigcup_{k\in\omega}\underline{A}_k} Y &\iff X \cup Y \subseteq \textstyle\bigcup_{k\in\omega} A_k \ \&\ \exists i.\ X \vdash_i Y \\ &\iff X \cup Y \subseteq \textstyle\bigcup_{k\in\omega} A_k \ \&\ X \vdash_{\underline{B}} Y. \end{aligned}$$

□

The substructure relation $\trianglelefteq$ can be extended to $(n+1)$-tuples of sequent structure coordinatewise. More precisely we require

$$(\underline{A}_0, \underline{A}_1, \cdots \underline{A}_n) \trianglelefteq (\underline{B}_0, \underline{B}_1, \cdots \underline{B}_n)$$

if and only if for each $0 \leq i \leq n$, $\underline{A}_i \trianglelefteq \underline{B}_i$. For convenience write $\vec{\underline{A}}$ for $(\underline{A}_0, \underline{A}_1, \cdots \underline{A}_n)$.

The least upper bound of an ω-increasing chain of n-tuples of sequent structures is then just the n-tuple of sequent structures consisting of the least upper bounds on each component, *i.e.* if

$$\vec{\underline{A}}_1 \trianglelefteq \vec{\underline{A}}_2 \cdots \trianglelefteq \vec{\underline{A}}_i \trianglelefteq \cdots$$

then

$$\pi_j \left(\bigsqcup_{i \in \omega} \vec{\underline{A}}_i \right) = \bigcup_{i \in \omega} \pi_j \left(\vec{\underline{A}}_i \right).$$

An operation F from n-tuples of sequent structures to m-tuples of sequent structures is said to be *continuous* if it is monotonic, *i.e.* $\vec{\underline{A}} \trianglelefteq \vec{\underline{B}}$ implies $F(\vec{\underline{A}}) \trianglelefteq F(\vec{\underline{B}})$ and it preserves ω-increasing chains of sequent structures, *i.e.*

$$\vec{\underline{A}}_1 \trianglelefteq \vec{\underline{A}}_2 \cdots \trianglelefteq \vec{\underline{A}}_i \trianglelefteq \cdots$$

implies

$$\bigcup_{i \in \omega} F(\vec{\underline{A}}_i) = F(\bigcup_{i \in \omega} \vec{\underline{A}}_i).$$

It is well known that for functions on tuples of cpos they are continuous if and only if by changing any argument while fixing others the induced function is continuous.

Larsen and Winskel [LW84] have a useful lemma which concludes that an operation F is continuous if and only if it is monotonic with respect to $\trianglelefteq$ and continuous on proposition sets in the sense that for any ω-increasing chain

$$\vec{\underline{A}}_1 \trianglelefteq \vec{\underline{A}}_2 \cdots \trianglelefteq \vec{\underline{A}}_i \trianglelefteq \cdots,$$

each proposition of $F(\bigcup_{i \in \omega} \vec{\underline{A}}_i)$ is a proposition of $\bigcup_{i \in \omega} F(\vec{\underline{A}}_i)$. Generalized to SFSSs we have

Lemma 3.1 *Let F be a function on* $\mathbf{CPO}_{\mathrm{SFSS}}$. *$F$ is continuous if and only if it is monotonic and continuous on proposition sets.*

We conclude this chapter by the following theorem, which implies the existence of recursively defined systems using constructions introduced in some previous sections.

Theorem 3.9 *$(\ )_{\uparrow}$, $+$, $\times$, $\rightarrow$, P_H, P_S, and P_P are all continuous.*

Proof We illustrate the proof for $\rightarrow$. Proofs for other cases follow the same pattern, hence omitted.

We have to show that $\rightarrow$ is a continuous operation from pairs of sequent structures to sequent structures. $\rightarrow$ is monotonic in its first argument. Suppose $\underline{A} \trianglelefteq \underline{A}'$ and $\underline{B}$ are sequent structures. Write

$$\underline{C} = (C, \vdash) = \underline{A} \rightarrow \underline{B}$$

and

$$\underline{C}' = (C', \vdash') = \underline{A}' \rightarrow \underline{B}.$$

We check 1 and 2 in Definition 3.19 to show that $\underline{C} \trianglelefteq \underline{C}'$.

1. Assume $X \in C$. It is easy to see that $X \in C'$.

2. Clearly $X \vdash_{\underline{C}} Y$ implies $X \vdash_{\underline{C}'} Y$. Assume $X \subseteq C$, $Y \subseteq C$ and $X \vdash_{\underline{C}'} Y$. Because in this case each entailment with subscript $\underline{A}'$ is an entailment with subscript $\underline{A}$, by the assumption that $\underline{A} \trianglelefteq \underline{A}'$. Hence we have $X \vdash_{\underline{C}} Y$.

Now we show that $\rightarrow$ is continuous on proposition sets. Let

$$\underline{A}_0 \trianglelefteq \underline{A}_1 \trianglelefteq \cdots \trianglelefteq \underline{A}_i \trianglelefteq \cdots$$

be a chain of SFSSs. Let X be a proposition of $(\bigcup_{i\in\omega} \underline{A}_i) \rightarrow \underline{B}$. Then $\pi_1 X \Subset \bigcup_{i\in\omega} A_i$. Hence $\pi_1 X \subseteq A_j$ for some j, which means X is a proposition of $\underline{A}_j \rightarrow \underline{B}$. Thus X is a proposition of $\bigcup_{i\in\omega}(\underline{A}_i \rightarrow \underline{B})$. By Lemma 3.1, $\rightarrow$ is continuous in its first argument. Similarly we can prove that $\rightarrow$ is continuous in its second argument, and hence continuous.

□

Chapter 4

A Logic of SFP

This chapter introduces a logic of **SFP**. The logical framework lays the foundation for the derivation, from a denotational semantics, a sound and complete proof system for a programming language. Given a programming language, the domain for its denotational semantics specifies a type. That type allows the identification, in a uniform manner, an assertion language and a proof system from the general framework. Any proof system so derived is guaranteed to agree with the denotational semantics in the sense described in Section 1.4.

This chapter has the following sections. Section 4.1 introduces a meta-language for denotational semantics with type constructions such as sum, product, function space, the three powerdomains, and recursively defined types. For each type there is an assertion language, constructed from the languages of the component types. Section 4.2 gives proof systems for the typed assertion languages. The proof rules axiomatize the entailment $\leq$ between assertions. They are formulated in a compositional manner. Section 4.3 interprets assertions as Scott open sets and proves the soundness of the proof system. Section 4.4 discusses completeness and expressiveness of the proof system. Section 4.5 deals with relative completeness. Section 4.6 presents an application of the framework, showing that the assertions Brookes proposed in [Br85] is determined by Plotkin's domain of resumptions [Pl76]. The final section of this chapter provides a proof system for non-entailment.

4.1 Typed Assertion Languages

Our framework consists of four parts: a meta-language for denotational semantics, typed assertion languages, structural and logical rules and meta-predicates. This section is concerned with the meta-language for denotational semantics and typed assertion languages.

The meta-language for denotational semantics is usually introduced as a language of type expressions as follows:

$$\sigma ::= \mathbf{1} \mid \sigma \times \tau \mid \sigma \to \tau \mid \sigma + \tau \mid \sigma_\bot \mid \mathcal{P}_S[\sigma] \mid \mathcal{P}_H[\sigma] \mid \mathcal{P}_P[\sigma] \mid t \mid rec\ t.\sigma,$$

where t is a type variable and σ, τ ranges over type expressions.

Note we could have avoided using $\mathbf{1}$ as a ground type because it can be expressed by the recursively defined type $rec\ t.t$. Every closed type expression can be interpreted as an SFP domain, with $\mathbf{1}$ as the one-point domain, $\times$ as the cartesian product, $+$ as the coalesced sum, $(\ )_\bot$ as lifting, $\to$ as the function space, $\mathcal{P}_S$, $\mathcal{P}_H$ and $\mathcal{P}_P$ as the Smyth, the Hoare, and the Plotkin powerdomains, and $rec\ t.\sigma$ the recursively defined domains. We can use the strongly finite sequent structures introduced in the previous chapter to give recursively defined domains. Write $\mathcal{D}(\sigma)$ for the domain corresponding to the type σ.

Using this meta-language we can give denotational semantics to most programming languages. For a programming language L, we specify a type expression σ and denote a program as an element in $\mathcal{D}(\sigma)$. Here we are not concerned with the problem of how the type expression is selected for a particular programming language.

For each type σ we introduce an assertion language $\mathcal{A}_\sigma$. The assertions of $\mathcal{A}_\sigma$ are constructed according to the way σ is built up from its subtypes. We use the notations for type constructions again in the assertion language to make a clear correspondence between types and typed assertions, just as we did for type constructions and constructions on domains.

In the following table for the syntax of assertions, rule (**t**,**f**) says that **t** and **f** are assertions of every type, standing for the logical 'true' and 'false'. Rule $(\wedge - \vee)$ means that assertions are closed under conjunction and disjunction. The rest of the rules tell us how assertions are built up according to the structure of the types. For example, according to $(\times)$, if φ is an assertion of a type σ and ψ is an assertion of a type τ, then

$\varphi \times \psi$ is an assertion of the type $\sigma \times \tau$. We do not need extra structures for recursively defined types other than what is shown in (*rec*). This is because, as shown in Chapter 3, it is possible to get equality in solving recursive domain equations.

In this way, given any type expression σ, an assertion language $\mathcal{A}_\sigma$ is provided.

Assertion Formation Rules

$(\mathbf{t},\mathbf{f})$ $\quad \mathbf{t},\mathbf{f} : \sigma$

$(\wedge - \vee)$ $\quad \dfrac{\varphi, \psi : \sigma}{\varphi \wedge \psi : \sigma \qquad \varphi \vee \psi : \sigma}$

$(\times)$ $\quad \dfrac{\varphi : \sigma \qquad \psi : \tau}{\varphi \times \psi : \sigma \times \tau}$

$(\rightarrow)$ $\quad \dfrac{\varphi : \sigma \qquad \psi : \tau}{\varphi \rightarrow \psi : \sigma \rightarrow \tau}$

$(+)$ $\quad \dfrac{\varphi : \sigma \qquad \psi : \tau}{\mathit{inl}\ \varphi : \sigma + \tau \qquad \mathit{inr}\ \psi : \sigma + \tau}$

$(\bot)$ $\quad \dfrac{\varphi : \sigma}{\varphi_\bot : \sigma_\bot}$

(H) $\quad \dfrac{\varphi : \sigma}{\Diamond\varphi : \mathcal{P}_H[\sigma]}$

(S) $\quad \dfrac{\varphi : \sigma}{\Box\varphi : \mathcal{P}_S[\sigma]}$

(P) $\quad \dfrac{\varphi : \sigma}{\Box\varphi : \mathcal{P}_P[\sigma],\ \Diamond\varphi : \mathcal{P}_P[\sigma]}$

(rec) $\quad \dfrac{\varphi : \sigma[\mathit{rec}\ t.\sigma/t]}{\varphi :\ \mathit{rec}\ t.\sigma}$

4.2 Proof Systems

The proof system for a type σ consists of rules for reasoning about entailments $\leq_\sigma$ between assertions of $\mathcal{A}_\sigma$. It is intended that if $\varphi \leq_\sigma \psi$ then every program satisfying φ also satisfies ψ. This intuition will be made precise later when we give an interpretation of assertions. The logical equivalence, $=_\sigma$, stands for the conjunction of $\leq_\sigma$ and $\geq_\sigma$. When $\varphi,\ \psi \in \mathcal{A}_\sigma$, $\varphi \leq_\sigma \psi$ and $\varphi =_\sigma \psi$ are called *formulae*.

The proof system consists of rules of the form

$$\frac{A_1 \quad A_2 \quad \cdots A_n}{B}$$

where A_i's and B are formulae. Such a rule means that if A_i's are valid, then so is B. In the special case where the assumption is empty, a rule becomes an axiom. For convenience let $\bigvee\emptyset \equiv \mathbf{f}$ and $\bigwedge\emptyset \equiv \mathbf{t}$.

For each type σ we have the following logical axioms and rules:

Logical Axioms and Rules

$$(\mathbf{t}) \quad \varphi \leq \mathbf{t} \qquad (\mathbf{f}) \quad \mathbf{f} \leq \varphi$$

$$(\text{Ref}) \quad \varphi \leq \varphi$$

$$(\text{Trans}) \quad \frac{\varphi \leq \varphi' \quad \varphi' \leq \varphi''}{\varphi \leq \varphi''}$$

$$(\leq - =) \quad \frac{\varphi \leq \psi \quad \psi \leq \varphi}{\varphi = \psi}$$

$$(= - \leq) \quad \frac{\varphi = \varphi'}{\varphi \leq \varphi'} \quad \frac{\varphi = \varphi'}{\varphi' \leq \varphi}$$

$$(\wedge - \leq) \quad \varphi \wedge \varphi' \leq \varphi \quad \varphi \wedge \varphi' \leq \varphi'$$

$$(\leq - \wedge) \quad \frac{\varphi \leq \varphi' \quad \varphi \leq \varphi''}{\varphi \leq \varphi' \wedge \varphi''}$$

$$(\vee - \leq) \quad \frac{\varphi \leq \varphi' \quad \psi \leq \varphi'}{\varphi \vee \psi \leq \varphi'}$$

$$(\leq - \vee) \quad \varphi \leq \varphi' \vee \varphi \quad \varphi' \leq \varphi' \vee \varphi$$

$$(\wedge - \vee) \quad \varphi \wedge (\varphi_1 \vee \varphi_2) \leq (\varphi \wedge \varphi_1) \vee (\varphi \wedge \varphi_2)$$

There are type-specific rules for reasoning about entailments of assertions of composite types. There are also axioms that tell us how logical constructions interact with type constructions. For product we have

Product

$$(\times - \mathbf{t}) \quad \mathbf{t}_\sigma \times \mathbf{t}_\tau =_{\sigma\times\tau} \mathbf{t}_{\sigma\times\tau}$$

$$(\times - \mathbf{f}) \quad \mathbf{f}_\sigma \times \varphi = \mathbf{f}_{\sigma\times\tau} \quad \varphi \times \mathbf{f}_\tau =_{\sigma\times\tau} \mathbf{f}_{\sigma\times\tau}$$

$$(\times - \vee) \quad (\psi_1 \vee \psi_2) \times (\varphi_1 \vee \varphi_2) =_{\sigma\times\tau} (\psi_1 \times \varphi_1) \vee (\psi_1 \times \varphi_2) \vee (\psi_2 \times \varphi_1) \vee (\psi_2 \times \varphi_2)$$

$$(\times - \wedge) \quad (\varphi_1 \times \varphi_2) \wedge (\psi_1 \times \psi_2) =_{\sigma\times\tau} (\varphi_1 \wedge \psi_1) \times (\varphi_2 \wedge \psi_2)$$

These axioms and rules are self-evident. Similarly, there are axioms and rules for other type constructions. For simplicity subscripts are omitted.

Sum

$(+-\mathbf{t})$	$\mathit{inl}\,\mathbf{t} = \mathbf{t} \quad \mathit{inr}\,\mathbf{t} = \mathbf{t}$
$(+-\mathbf{f})$	$\mathit{inl}\,\mathbf{f} = \mathbf{f} \quad \mathit{inr}\,\mathbf{f} = \mathbf{f}$
$(\mathit{inl} - \wedge)$	$\mathit{inl}\,(\varphi \wedge \psi) = (\mathit{inl}\,\varphi) \wedge (\mathit{inl}\,\psi)$
$(\mathit{inr} - \wedge)$	$\mathit{inr}\,(\varphi \wedge \psi) = (\mathit{inr}\,\varphi) \wedge (\mathit{inr}\,\psi)$
$(\mathit{inl} - \vee)$	$\mathit{inl}\,(\varphi \vee \psi) = (\mathit{inl}\,\varphi) \vee (\mathit{inl}\,\psi)$
$(\mathit{inr} - \vee)$	$\mathit{inr}\,(\varphi \vee \psi) = (\mathit{inr}\,\varphi) \vee (\mathit{inr}\,\psi)$

The intuitive reason for axioms like $(+-\mathbf{t})$ can be seen from the picture of the sum of two cpos.

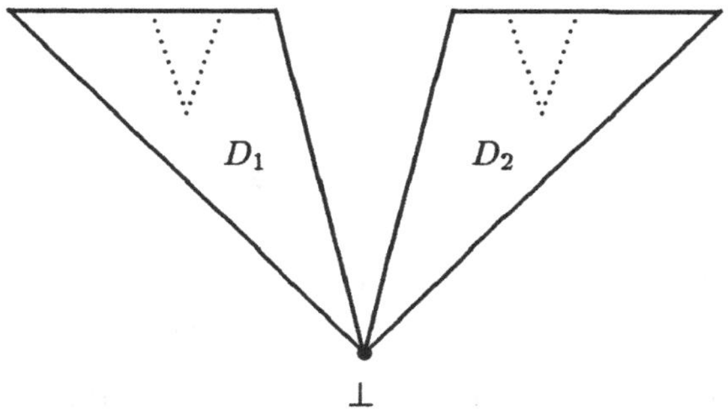

The upper parts of the dotted lines represent open sets. However, $\mathbf{t}$ in D_1 represents an open set which has the bottom $\perp_{D_1}$ in it. Considered as part of the sum $D_1 + D_2$, that open set must be the whole $D_1 + D_2$, since an open set must be upwards closed, and $D_1 + D_2$, D_1 share the same bottom.

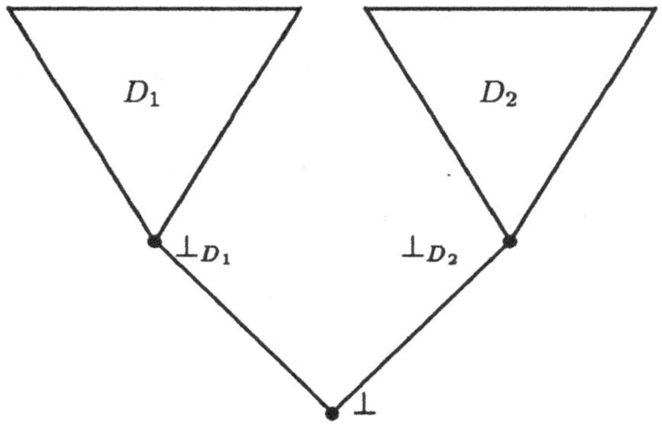

The same axiom $(+ - \mathbf{t})$ would not be valid for separated sum. Again, this can be best explained by a picture. We can see from the previous picture that when D_1 is considered as part of the separated sum, it is different from $(D_1)_\perp + (D_2)_\perp$ since this time a new bottom is introduced, different from that of D_1 and D_2. Here are rules for other types.

Lifting

$(\perp - \mathbf{f})$	$(\mathbf{f}_\sigma)_\perp = \mathbf{f}_{\sigma_\perp}$
$(\perp - \wedge)$	$(\varphi_1 \wedge \varphi_2)_\perp = (\varphi_1)_\perp \wedge (\varphi_2)_\perp$
$(\perp - \vee)$	$(\varphi_1 \vee \varphi_2)_\perp = (\varphi_1)_\perp \vee (\varphi_2)_\perp$

Function Space

$(\rightarrow - \mathbf{t})$	$\varphi \rightarrow \mathbf{t} = \mathbf{t} \qquad \mathbf{f} \rightarrow \varphi = \mathbf{t}$
$(\rightarrow - \wedge)$	$\varphi \rightarrow (\psi_1 \wedge \psi_2) = (\varphi \rightarrow \psi_1) \wedge (\varphi \rightarrow \psi_2)$
$(\vee - \rightarrow)$	$(\varphi_1 \vee \varphi_2) \rightarrow \psi = (\varphi_1 \rightarrow \psi) \wedge (\varphi_2 \rightarrow \psi)$

The Modality $\Box$

$(\Box - \mathbf{t})$	$\Box\mathbf{f} = \mathbf{f}$
$(\Box - \mathbf{f})$	$\Box\mathbf{t} = \mathbf{t}$
$(\Box - \wedge)$	$\Box(\varphi_1 \wedge \varphi_2) = (\Box\varphi_1) \wedge (\Box\varphi_2)$

Note axiom $(\Box - \mathbf{f})$ corresponds to the choice that the empty set is excluded from the Smyth powerdomain.

The Modality $\Diamond$

$(\Diamond - \mathbf{t})$	$\Diamond\mathbf{f} = \mathbf{f}$
$(\Diamond - \mathbf{f})$	$\Diamond\mathbf{t} = \mathbf{t}$
$(\Diamond - \vee)$	$\Diamond(\varphi_1 \vee \varphi_2) = (\Diamond\varphi_1) \vee (\Diamond\varphi_2)$

We remark that the axioms and rules just presented with respect to $\Box$ and $\Diamond$ are for the Smyth powerdomain and the Hoare powerdomain, respectively. They are used for the Plotkin powerdomain, too. However, the following two axioms are for the Plotkin powerdomain only.

$\Diamond - \Box$

$(\Diamond)$	$\Diamond\varphi \wedge \Box\psi \leq \Diamond(\varphi \wedge \psi)$
$(\Box)$	$\Box(\varphi \vee \psi) \leq \Diamond\varphi \vee \Box\psi$

Finally there is the *Context* rule which says that substituting an assertion by an equivalent one in any context preserves equivalence:

$$(\text{Context}) \qquad \frac{\varphi =_\sigma \psi}{C[\varphi] =_\tau C[\psi]}.$$

Here by a context $C[\ \ ]$ we mean an assertion with holes in it, and these holes can be filled in by appropriate assertions. Examples of context include: $[\ \ ]$, $\varphi \times [\ \ ] \vee u \times v$, and $([\ \ ] \to \varphi) \wedge ([\ \ ] \to \psi)$, and so on.

Many new rules can now be derived from the combination of type-specific rules, logical rules, and Context. For example, a rule

$$\frac{\psi' \leq \psi \quad \varphi \leq \varphi'}{\psi \to \varphi \leq \psi' \to \varphi'}$$

for function space can be derived in the following way. Suppose $\varphi \leq \varphi'$ and $\psi' \leq \psi$. We have $\varphi = \varphi \wedge \varphi'$ and $\psi = \psi \vee \psi'$. Hence

$$\begin{aligned}
\psi \to \varphi &= \psi \to \varphi \wedge \varphi' && (\mathit{Contx}) \\
&= (\psi \to \varphi) \wedge (\psi \to \varphi') && (\to - \wedge) \\
&\leq \psi \to \varphi' && \text{logical} \\
&= \psi \vee \psi' \to \varphi' && (\mathit{Contx}) \\
&= (\psi' \to \varphi') \wedge (\psi \to \varphi') && (\vee - \to) \\
&\leq \psi' \to \varphi'. && \text{logical}
\end{aligned}$$

Here is the collection of some useful derived rules:

$$(\times - \leq) \qquad \frac{\psi \leq \psi' \quad \varphi \leq \varphi'}{\psi \times \varphi \leq \psi' \times \varphi'}$$

$$(\mathit{inl} - \leq) \qquad \frac{\varphi \leq \psi}{\mathit{inl}\,\varphi \leq \mathit{inl}\,\psi} \qquad\qquad (\mathit{inr} - \leq) \qquad \frac{\varphi \leq \psi}{\mathit{inr}\,\varphi \leq \mathit{inr}\,\psi}$$

$$(\bot - \leq) \qquad \frac{\varphi \leq \psi}{\varphi_\bot \leq \psi_\bot}$$

$$(\to - \leq) \qquad \frac{\psi' \leq \psi \quad \varphi \leq \varphi'}{\psi \to \varphi \leq \psi' \to \varphi'}$$

$$(\Box - \leq) \qquad \frac{\varphi \leq \psi}{\Box\varphi \leq \Box\psi}$$

$$(\Diamond - \leq) \qquad \frac{\varphi \leq \psi}{\Diamond\varphi \leq \Diamond\psi}$$

For the syntax of assertions we remark that type constructions $\times$, $\to$, *inl*, *inr*, $(\ \ )_\bot$, $\Diamond$, and $\Box$ are given priority over logical operations. For example, when writting $\mathit{inl}\,\varphi \vee \mathit{inr}\,\psi$ we mean $(\mathit{inl}\,\varphi) \vee (\mathit{inr}\,\psi)$.

4.3 Soundness

For each closed type expression σ we define an *interpretation function*

$$[\![\ ]\!]_\sigma : \mathcal{A}_\sigma \to \mathbf{K}\Omega(\mathcal{D}(\sigma))$$

in the following structural way:

$$\begin{aligned}
[\![\mathbf{t}]\!]_\sigma &= \mathcal{D}(\sigma),\\
[\![\mathbf{f}]\!]_\sigma &= \emptyset,\\
[\![\varphi \vee \psi]\!]_\sigma &= [\![\varphi]\!]_\sigma \cup [\![\psi]\!]_\sigma,\\
[\![\varphi \wedge \psi]\!]_\sigma &= [\![\varphi]\!]_\sigma \cap [\![\psi]\!]_\sigma.
\end{aligned}$$

With respect to type constructions we define

$$\begin{aligned}
[\![\varphi \times \psi]\!]_{\sigma\times\tau} &= \{\langle u, v\rangle \mid u \in [\![\varphi]\!]_\sigma \ \&\ v \in [\![\psi]\!]_\tau\},\\
[\![\mathit{inl}\,\varphi]\!]_{\sigma+\tau} &= \{\langle 0, u\rangle \mid u \in [\![\varphi]\!]_\sigma \setminus \{\bot_{\mathcal{D}(\sigma)}\}\} \cup\\
&\qquad \{x \in \mathcal{D}(\sigma+\tau) \mid \bot_{\mathcal{D}(\sigma)} \in [\![\varphi]\!]_\sigma\},\\
[\![\mathit{inr}\,\varphi]\!]_{\sigma+\tau} &= \{\langle 1, u\rangle \mid u \in [\![\varphi]\!]_\tau \setminus \{\bot_{\mathcal{D}(\tau)}\}\} \cup\\
&\qquad \{x \in \mathcal{D}(\sigma+\tau) \mid \bot_{\mathcal{D}(\tau)} \in [\![\varphi]\!]_\tau\},\\
[\![\varphi \to \psi]\!]_{\sigma\to\tau} &= \{f \in [\mathcal{D}(\sigma) \to \mathcal{D}(\tau)] \mid [\![\varphi]\!]_\sigma \subseteq f^{-1}([\![\psi]\!]_\tau)\},\\
[\![(\varphi)_\bot]\!]_{(\sigma)_\bot} &= \{\langle 0, u\rangle \mid u \in [\![\varphi]\!]_\sigma\},\\
[\![\Box\varphi]\!]_{\mathcal{P}_S(\sigma)} &= \{U \in \mathcal{P}_S(\mathcal{D}\sigma) \mid U \subseteq [\![\varphi]\!]_\sigma\},\\
[\![\Box\varphi]\!]_{\mathcal{P}_P(\sigma)} &= \{U \in \mathcal{P}_P(\mathcal{D}\sigma) \mid U \subseteq [\![\varphi]\!]_\sigma\},\\
[\![\Diamond\varphi]\!]_{\mathcal{P}_H(\sigma)} &= \{L \in \mathcal{P}_H(\mathcal{D}\sigma) \mid L \cap [\![\varphi]\!]_\sigma \neq \emptyset\},\\
[\![\Diamond\varphi]\!]_{\mathcal{P}_P(\sigma)} &= \{L \in \mathcal{P}_P(\mathcal{D}\sigma) \mid L \cap [\![\varphi]\!]_\sigma \neq \emptyset\},\\
[\![\varphi]\!]_{rect.\sigma} &= \{\epsilon_\sigma(u) \mid u \in [\![\varphi]\!]_{\sigma[(rect.\sigma)\backslash t]}\},
\end{aligned}$$

where $\epsilon_\sigma : [\,\mathcal{D}(\sigma[(rect.\sigma)\backslash t]) \to \mathcal{D}(rect.\sigma)\,]$ is the isomorphism arising form the recursively defined domain for the type $rec\,t.\,\sigma$.

Definition 4.1 *For assertions φ, ψ of $\mathcal{A}_\sigma$, write $\models_\sigma \varphi \leq_\sigma \psi$ if $[\![\varphi]\!]_\sigma \subseteq [\![\psi]\!]_\sigma$. Call a formula $\varphi \leq_\sigma \psi$ valid if $\models_\sigma \varphi \leq_\sigma \psi$. Write $\vdash_\sigma \varphi \leq_\sigma \psi$ if $\varphi \leq_\sigma \psi$ can be derived from axioms and rules given in Section 4.2 together with those presented in Section 4.4 later. A rule is sound if it produces valid formulae from valid formulae. The proof system is called sound if $\vdash \varphi \leq_\sigma \psi$ implies $\models \varphi \leq_\sigma \psi$. It is complete if $\models \varphi \leq_\sigma \psi$ implies $\vdash \varphi \leq_\sigma \psi$.*

It is clear that a proof system is sound if and only if all its axioms are valid and rules sound. The next proposition establishes the soundness of the proof system.

Proposition 4.1

- *The logical axioms are valid and logical rules sound.*
- *The axioms for product, sum, function space, lifting are valid.*
- *The rules for product, sum, function space, lifting are sound.*
- *The axioms associated with* $\Box$, $\Diamond$ *are valid and rules sound.*

Proof As an illustration we show the validness of the $\Box - \Diamond$ axioms. Let A, B be compact open sets of an SFP domain D. Clearly

$$T \in \{U \in \mathcal{P}_P(D) \mid U \cap A \neq \emptyset\} \cap \{V \in \mathcal{P}_P(D) \mid V \subseteq B\}$$
$$\Longrightarrow T \in \{U \in \mathcal{P}_P(D) \mid U \cap A \cap B \neq \emptyset\}.$$

Hence

$$\{U \in \mathcal{P}_P(D) \mid U \cap A \neq \emptyset\} \cap \{V \in \mathcal{P}_P(D) \mid V \subseteq B\}$$
$$\subseteq \{U \in \mathcal{P}_P(D) \mid U \cap A \cap B \neq \emptyset\}.$$

So ($\Diamond$) is valid. ($\Box$) is valid because

$$T \in \{U \in \mathcal{P}_P(D) \mid U \subseteq A \cup B\}$$
$$\Longrightarrow T \in \{U \in \mathcal{P}_P(D) \mid U \subseteq B\} \cup \{U \in \mathcal{P}_P(D) \mid U \cap A \neq \emptyset\}.$$

The rest of the proof can be similarly carried out by inspecting each axiom and rule.

$\Box$

In many proofs of this section and next section we will not check the case for recursively defined types, because all the time we are dealing with finite sets of assertions, and these assertions can always be considered as of some finite subtype of the recursively defined type.

4.4 Completeness

Note that the proof system given in Section 4.2 is not complete since there are some valid formulae which are not derivable, as shown by the following example.

Example 4.1 *Consider the type* $\mathbf{1} \to [(\mathbf{1}_\perp) + (\mathbf{1}_\perp)]$ *and the formula*

$$\mathbf{t} \to (\mathit{inr}(\mathbf{t}_\perp) \vee \mathit{inl}(\mathbf{t}_\perp)) = [\mathbf{t} \to \mathit{inr}(\mathbf{t}_\perp)] \vee [\mathbf{t} \to \mathit{inl}(\mathbf{t}_\perp)]$$

of this type.

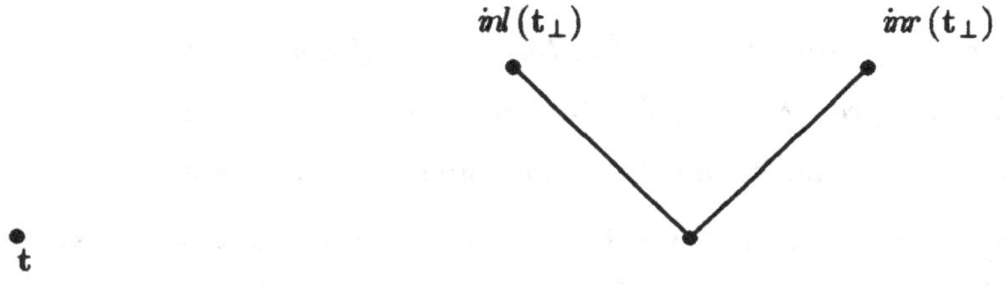

$\mathcal{D}(\mathbf{1})$ $\qquad\qquad$ $\mathcal{D}(\mathbf{1}_\perp + \mathbf{1}_\perp)$

This formula says that a function maps the bottom of the first domain to either $\mathit{inl}(\mathbf{t}_\perp)$ *or* $\mathit{inr}(\mathbf{t}_\perp)$ *of the second domain if and only if it either maps the bottom to* $\mathit{inl}(\mathbf{t}_\perp)$*, or maps it to* $\mathit{inr}(\mathbf{t}_\perp)$*. It is obviously valid but it cannot be derived from the proof system given in Section 4.2, because there is no rule for manipulating assertions of the form*

$$\varphi \to (\alpha \vee \beta).$$

Motivated from this example, we introduce further proof rules to get a complete proof system. One axiom to be introduced would be

$$\varphi \to \bigvee_{i \in I} \alpha_i = \bigvee_{i \in I} \varphi \to \alpha_i,$$

but clearly it is not valid. However, it becomes valid when we require φ to be some restricted assertions, those which are *prime*. For that purpose a meta-predicate is needed. Prime assertions are captured by the predicate **P**.

When $\mathbf{P}(\varphi)$, as specified in the table next page, we say φ is a prime assertion. According to this definition $\mathit{inl}\,\varphi \wedge \mathit{inr}\,\psi$, for example, is not a prime assertion even when both φ and ψ are. Similarly $\Box\varphi \wedge \Box\psi$ is not a prime assertion either. It is worth noticing that whether an assertion is prime or not is purely a syntactical question. Also note that the interpretation $[\![\varphi]\!]$ being a prime open set does not imply that φ is a prime assertion. On the other hand, however, to get completeness we must make sure that for every prime open set O, there is a prime assertion ψ such that $[\![\psi]\!] = O$. It is indeed the case, and will be checked later.

Prime Assertions

$$(\mathbf{t}) \quad \mathbf{P}(\mathbf{t})$$

$$(\times) \quad \frac{\mathbf{P}(\varphi) \quad \mathbf{P}(\psi)}{\mathbf{P}(\varphi \times \psi)}$$

$$(+) \quad \frac{\mathbf{P}(\varphi)}{\mathbf{P}(\mathit{inl}\,\varphi)} \qquad \frac{\mathbf{P}(\psi)}{\mathbf{P}(\mathit{inr}\,\psi)}$$

$$(\bot) \quad \frac{\mathbf{P}(\varphi)}{\mathbf{P}(\varphi_\bot)}$$

$$(\rightarrow) \quad \frac{\begin{array}{ll} \forall i \in I.\ \mathbf{P}(\varphi_i)\ \&\ \mathbf{P}(\psi_i) & \& \\ \forall I' \subseteq I \exists I'' \subseteq I \exists J. & (\bigwedge_{i \in I'} \varphi_i = \bigvee_{i \in I''} \varphi_i)\ \& \\ & (\bigwedge_{i \in I'} \psi_i = \bigvee_{j \in J} \chi_j)\ \& \\ & (\forall i \in I'' \exists j \in J.\ \varphi_i \equiv \chi_j) \end{array}}{\mathbf{P}(\bigwedge_{i \in I} \varphi_i \rightarrow \psi_i)}$$

$$(H) \quad \frac{\mathbf{P}(\varphi_i) \quad 1 \leq i \leq n}{\mathbf{P}(\bigwedge_{1 \leq i \leq n} \Diamond \varphi_i)}$$

$$(S) \quad \frac{\mathbf{P}(\varphi_i) \quad 1 \leq i \leq n}{\mathbf{P}(\Box(\bigvee_{1 \leq i \leq n} \varphi_i))}$$

$$(P) \quad \frac{\mathbf{P}(\varphi_i) \quad 1 \leq i \leq n}{\mathbf{P}([\Box \bigvee_{1 \leq i \leq n} \varphi_i] \wedge \bigwedge_{1 \leq i \leq n} \Diamond \varphi_i)}$$

Particularly worth noticing is perhaps the rule $(\rightarrow)$ for function space, where $\equiv$ stands for syntaxtic equality. Careful readers may have observed that this rule bears great resemblance to Definition 2.1 as well as Definition 3.9. The reader may have also noticed that the second existential quantification $\exists J$ is unbounded. However, one does not have to search the whole collection of assertions to find the χ's. Later we will see that there is a strategy to transform $\bigwedge_{i \in I'} \psi_i$ into an equivalent prime normal form. This should give us enough hint for finding the χ's effectively and thus makes the predicate easy to use.

Note that prime assertions of one type may not be that of another for powerdomains. In other words, we use (H) for the Hoare powerdomain, (S) for the Smyth powerdomain, and (P) for the Plotkin powerdomain, respectively.

Now we are in a position to finish the presentation of the proof system by introducing two more rules:

$$(\rightarrow -\vee) \qquad \frac{\mathbf{P}(\varphi)}{\varphi \rightarrow \bigvee_{i\in I} \alpha_i = \bigvee_{i\in I} \varphi \rightarrow \alpha_i}$$

$$(\mathit{inl} - \mathit{inr}) \qquad \frac{\mathbf{Q}(\varphi) \qquad \mathbf{Q}(\psi)}{\mathit{inl}\,\varphi \wedge \mathit{inr}\,\psi \leq \mathbf{f}}$$

Here $\mathbf{Q}$ is a much simpler predicate on assertions than $\mathbf{P}$. For a prime assertion φ, $\mathbf{Q}(\varphi)$ *if in its syntax every appearance of* $\mathbf{t}$ *occurs under some context* $(\quad)_\perp$ *except for assertions of the form* $\varphi\times\mathbf{t}$, $\mathbf{t}\times\varphi$, *and* $\mathbf{t} \rightarrow \varphi$ *where only* φ *is required to have this property.* For example, we have $\mathbf{Q}(\mathbf{t} \times \mathbf{t}_\perp)$, but not $\Diamond\mathbf{t}$. We do not have $\mathbf{Q}(\mathbf{f} \rightarrow \mathbf{t}_\perp)$, since $\mathbf{f} \rightarrow \mathbf{t}_\perp$ is not a prime assertion (because $\mathbf{f}$ is not). The predicate $\mathbf{Q}$ is intended to capture some prime assertions which are not equivalent to $\mathbf{t}$. In other words, if $\mathbf{Q}(\varphi)$, it should be the case that $\perp \notin [\![\varphi]\!]$. The reverse, however, that if $\perp \notin [\![\varphi]\!]$ then $\mathbf{Q}(\varphi)$ is clearly not true (and clearly not needed). For the purpose of completeness all we need is that for every prime assertion ψ, if $\perp \notin [\![\psi]\!]$, then there is some φ such that $\varphi = \psi$, $\mathbf{Q}(\varphi)$, and $\perp \notin [\![\varphi]\!]$.

If for an assertion φ, $[\![\varphi]\!]$ is a prime open set then there is a finite element a such that $[\![\varphi]\!] = \uparrow a$. For notational convenience, write $\check{\varphi}$ for such an a.

Theorem 4.1 *For any closed type expression* σ *and assertion* $\varphi : \sigma$, $\mathbf{P}(\varphi)$ *implies that* $[\![\varphi]\!]_\sigma$ *is a (non-empty) prime open set, and* $\mathbf{Q}(\varphi)$ *implies that* $\perp \notin [\![\varphi]\!]_\sigma$.

Proof Straightforward by structural induction.

□

The proof system is sound as asserted by Proposition 4.1. However, after introduced the predicates $\mathbf{P}$, $\mathbf{Q}$, and the two new rules, the proof system remains sound. This is because the two new rules $(\rightarrow -\vee)$ and $(\mathit{inl} - \mathit{inr})$ are sound by an easy inspection in light of Theorem 4.1. Now we turn to the proof of completeness.

Definition 4.2 *Write* $\mathcal{P}_\sigma$ *for the proof system associated with type* σ. $\mathcal{P}_\sigma$ *is called prime complete if it has property* p_0, *prime normal if it has property*

p_1, and complete if it has property p_2, where

$$
\begin{array}{ll}
(p_0) & \forall\varphi,\ \psi:\sigma.\,(\mathbf{P}(\varphi)\ \&\ \mathbf{P}(\psi)\ \&\ [\![\varphi]\!]_\sigma \subseteq [\![\psi]\!]_\sigma \Longrightarrow \vdash \varphi \leq \psi),\\
(p_1) & \varphi:\sigma \Longrightarrow \exists\{\,\varphi_i \mid i\in I\,\}.\,\forall i\in I.\,\mathbf{P}(\varphi_i)\ \&\ \vdash \varphi = \bigvee_{i\in I}\varphi_i,\\
(p_2) & \forall\varphi,\ \psi:\sigma.\,([\![\varphi]\!] \subseteq [\![\psi]\!] \Longrightarrow \vdash \varphi \leq \psi).
\end{array}
$$

In (p_1), $\bigvee_{i\in I}\varphi_i$ is called a prime normal form of φ.

Clearly $\mathcal{P}_1$ has property p_0, p_1, and p_2. The proof of completeness is achieved by showing that each type construction preserves property (p_0), (p_1), and (p_2) in several propositions. Note that for each case the proof of (p_2) follows directly from (p_0) and (p_2). Given $\varphi,\ \psi:\sigma$, we can first reduce them to prime normal forms by (p_1) and have $\vdash \varphi = \bigvee_{i\in I}\varphi_i$ and $\vdash \psi = \bigvee_{j\in J}\psi_j$, where φ_i's and ψ_i's are prime assertions. We then have

$$
\begin{array}{lll}
\models \varphi \leq \psi & \Longrightarrow [\![\bigvee_{i\in I}\varphi_i]\!] \subseteq [\![\bigvee_{j\in J}\psi_j]\!] & \text{(Definition)}\\
& \Longrightarrow \forall i\,.\exists j.\ [\![\varphi_i]\!] \subseteq [\![\psi_j]\!] & \\
& \Longrightarrow \forall i\exists j.\ \vdash \varphi_i \leq \psi_j & (p_0)\\
& \Longrightarrow \forall i.\ \vdash \varphi_i \leq \bigvee_{j\in J}\psi_j & (\leq -\vee)\\
& \Longrightarrow \vdash \bigvee_{i\in I}\varphi_i \leq \bigvee_{j\in J}\psi_j. & (\vee- \leq)
\end{array}
$$

The following formula, obviously derivable from logical axioms and rules, will be frequently used for exchanging positions of $\bigvee$ and $\bigwedge$.

$$
(Ex)\qquad \bigwedge_{i=1}^{n}\bigvee_{j=1}^{n_i}\varphi_{i,j} = \bigvee_{f\in K}\bigwedge_{i=1}^{n}\varphi_{i,f(i)}
$$

with

$$
K = \{\, f:\overline{n} \rightarrow \overline{\max\{\, n_i \mid 1\leq i\leq n\,\}} \mid \forall i.\ f(i)\in\overline{n_i}\,\}
$$

where $\overline{s} = \{\,1,\ 2,\ \cdots,\ s\,\}$.

Proposition 4.2 *Product, sum, and lifting preserve property p_0, p_1, and p_2.*

The proofs are straightforward. However, we need some preparations first before showing that function space preserves property p_0, p_1, and p_2.

Definition 4.3 *Let $\{\varphi_i \mid i\in I\}$ be a finite set of assertions of type σ. It is said to be quasi-conjunctive closed if for every non-null $J\subseteq I$ there is a $K\subseteq I$ such that*

$$
\vdash \bigwedge_{j\in J}\varphi_j = \bigvee_{k\in K}\varphi_k.
$$

Since types are interpreted as SFP domains and assertions as open sets, it is not surprising that every finite set of assertions has a quasi-conjunctively closed super set in light of the results about SFP domains given in Chapter 2. More than that, we have the following rules for the construction of quasi-conjunctive closed supper set. The intended meaning of the operator Π^* is that for a finite set A of prime assertions the set $\Pi^*(A)$ is a quasi-conjunctive closed supper set of A. It is important to keep in mind that all the assertions appeared in the following rules are assumed to be prime.

Π^* Rules

$$(\mathbf{t}) \quad \Pi^*_\sigma(\{\mathbf{t}\}) = \{\mathbf{t}\}$$

$$(\bot) \quad \frac{\Pi^*_\sigma(\{\varphi \mid (\varphi)_\bot \in A\}) = P}{\Pi^*_{\sigma_\bot} A = \{(\varphi)_\bot \mid \varphi \in P\}}$$

$$(\times) \quad \frac{\Pi^*_\sigma(\{\varphi \mid \exists\psi.\varphi \times \psi \in A\}) = P \quad \Pi^*_\tau(\{\psi \mid \exists\varphi.\varphi \times \psi \in A\}) = Q}{\Pi^*_{\sigma\times\tau}(A) = \{\varphi \times \psi \mid \varphi \in P \ \&\ \psi \in Q\}}$$

$$(+) \quad \frac{\Pi^*_\sigma(\{\varphi \mid \mathit{inl}\,\varphi \in A\}) = U \quad \Pi^*_\tau(\{\varphi \mid \mathit{inr}\,\varphi \in A\}) = V}{\Pi^*_{\sigma+\tau}(A) = \{\mathit{inl}\,\varphi \mid \varphi \in U\} \cup \{\mathit{inr}\,\varphi \mid \varphi \in V\}}$$

$$(\rightarrow) \quad \frac{\Pi^*_\sigma(\bigcup_{i\in I} \pi_1\alpha_i) = A' \quad \Pi^*_\tau(\bigcup_{i\in I} \pi_2\alpha_i) = B'}{\Pi^*_{\sigma\to\tau}\{\alpha_i \mid i \in I\} = \{\beta \mid \mathbf{P}(\beta) \ \&\ \pi_1\beta \subseteq A' \ \&\ \pi_2\beta \subseteq B'\}}$$

$$(\Diamond) \quad \Pi^*_{\mathcal{P}_H(\sigma)}(A) = \{\bigwedge\Phi \mid \Phi \subseteq A\}$$

$$(\Box) \quad \frac{\Pi^*_\sigma(\bigcup\{\{\varphi_i \mid i \in I\} \mid \Box\bigvee_{i\in I}\varphi_i \in A\}) = T}{\Pi^*_{\mathcal{P}_S(\sigma)}(A) = \{\Box\bigvee\Phi \mid \Phi \subseteq T\}}$$

$$(\mathbf{P}) \quad \frac{\Pi^*_\sigma\{\varphi \mid \exists\alpha \in A.\,(\alpha \equiv \Box\bigvee\Phi \wedge \bigwedge_{\psi\in\Phi}\Diamond\psi \ \&\ \varphi \in \Phi)\} = S}{\Pi^*_{\mathcal{P}_P(\sigma)}(A) = \{\Box\bigvee\Phi \wedge \bigwedge_{\psi\in\Phi}\Diamond\psi \mid \Phi \subseteq S\}}$$

These rules provide a construction of quasi-conjunctively closed sets in a compositional way. We do not give a separate proof that these rules indeed produce quasi-conjunctively closed sets here. Instead, we point to places where the soundness of these rules is dealt with in principle. Axiom (**t**) is clearly sound. The soundness of ($\bot$), ($\times$), and ($+$) are treated in the proof of Theorem 3.7. The rules for powerdomains are discussed in the proofs of Proposition 3.8, Proposition 3.10, and Proposition 3.12. Finally, function space is dealt with in the proof of the following proposition.

Proposition 4.3 *Function space preserves property p_0, p_1, and p_2.*

Proof (p_0): We have

$$
\begin{aligned}
& [\![\bigwedge_{1\le i\le n} \varphi_i \to \psi_i]\!]_{\sigma\to\tau} \subseteq [\![\bigwedge_{1\le j\le m} \varphi'_j \to \psi'_j]\!]_{\sigma\to\tau} \\
\Longrightarrow & \bigcap_{1\le i\le n} [\![\varphi_i \to \psi_i]\!]_{\sigma\to\tau} \subseteq \bigcap_{1\le j\le m} [\![\varphi'_j \to \psi'_j]\!]_{\sigma\to\tau} \\
\Longrightarrow & \forall j. \bigcap_{1\le i\le n} [\![\varphi_i \to \psi_i]\!]_{\sigma\to\tau} \subseteq [\![\varphi'_j \to \psi'_j]\!]_{\sigma\to\tau} \\
\Longrightarrow & \forall j. \bigcap_{1\le i\le n} [\uparrow \check{\varphi}_i \to \uparrow \check{\psi}_i] \subseteq [\uparrow \check{\varphi}'_j \to \uparrow \check{\psi}'_j] \\
\Longrightarrow & \forall j. \bigsqcup_{1\le i\le n} [\check{\varphi}_i, \check{\psi}_i] \sqsupseteq [\check{\varphi}'_j, \check{\psi}'_j] \\
\Longrightarrow & \forall j\, \exists i.\ \check{\psi}_i \sqsupseteq \check{\psi}'_j \ \&\ \check{\varphi}_i \sqsubseteq \check{\varphi}'_i \\
\Longrightarrow & \forall j\, \exists i.\ \vdash \psi_i \le \psi'_j \ \&\ \vdash \varphi_i \ge \varphi'_j && \text{(by assumption)} \\
\Longrightarrow & \forall j\, \exists i.\ \vdash \varphi_i \to \psi_i \le \varphi'_j \to \psi'_j && (\to - \le) \\
\Longrightarrow & \forall j.\ \vdash \bigwedge_{1\le i\le n} \varphi_i \to \psi_i \ \le \varphi'_j \to \psi'_j \\
\Longrightarrow & \vdash \bigwedge_{1\le i\le n} \varphi_i \to \psi_i \le \bigwedge_{1\le j\le m} \varphi'_j \to \psi'_j .
\end{aligned}
$$

(p_1): We outline the steps leading to the prime normal form. Given any assertion $\theta : \sigma \to \tau$, we first use rules $(\to -\vee)$, $(\to -\wedge)$, $(\vee- \to)$ as well as (Ex) to put it in the following form:

$$\bigvee_{i=1}^{n} \bigwedge_{j=1}^{m_i} (\varphi_{i,j} \to \psi_{i,j})$$

where $\varphi_{i,j}, \psi_{i,j}$ are prime assertions of σ and τ, respectively. It is then enough to reduce each assertion of the form $\bigwedge_{i\in I}(\varphi_i \to \psi_i)$ to a finite disjunction of prime assertions.

Note that by the rules $(\to - \le)$ and $(\to -\wedge)$ for the function space, we have, for any $J \subseteq I$,

$$\vdash \bigwedge_{i\in I}(\varphi_i \to \psi_i) = (\bigwedge_{j\in J} \varphi_j \to \bigwedge_{j\in J} \psi_j) \wedge \bigwedge_{i\in I}(\varphi_i \to \psi_i).$$

Now we can substitute $\bigwedge_{j\in J} \varphi_j$ and $\bigwedge_{j\in J} \psi_j$ by their normal forms (which are assumed to exist) and apply $(\to -\vee)$ and $(\vee- \to)$ to put the assertion back into the form

$$\bigwedge_{i\in I}(\varphi_i \to \psi_i).$$

By repeating the procedure, we have, at the end,

$$\begin{aligned} \vdash & \bigwedge_{i \in I} (\varphi_i \to \psi_i) \\ & = \bigvee_k (\bigwedge_{i \in P} \alpha_i \to \beta_{i\,k}) \end{aligned}$$

where each $\bigwedge_{i \in P} \alpha_i \to \beta_{i\,k}$ is a prime assertion with

$$\Pi^*\{ \varphi_i \mid i \in I \} = \{ \alpha_k \mid k \in P \},$$

$$\Pi^*\{ \psi_i \mid i \in I \} = \{ \beta_k \mid k \in Q \},$$

and $\beta_{i\,k}$'s from $\{ \beta_k \mid k \in Q \}$.

□

The Hoare and the Smyth powerdomain are easy to deal with and hence the proof is omitted.

Proposition 4.4 *Hoare and Smyth powerdomains preserve property* p_0, p_1, *and* p_2.

That the Plotkin powerdomain preserves (p_1) is more complicated to prove. By exchanging the position of $\bigvee$ and $\bigwedge$ we can reduce an assertion into a disjunction of conjunction of assertions $\Box \bigvee_i \varphi_i \wedge \bigwedge_j \Diamond\psi_j$, where φ_i's and ψ_j's are prime. We need to show that each $\Box \bigvee_i \varphi_i \wedge \bigwedge_j \Diamond\psi_j$ is equivalent to a disjunction of prime assertions. Our idea is to show that, first, any $\Box \bigvee_i \varphi_i$ is equivalent to a disjunction of prime assertions. We then show that if θ is prime, then $\theta \wedge \Diamond\psi$ is equivalent to a disjunction of prime assertions. Applying this result as many times as necessary, we can 'absorb' any $\Diamond\psi$ in $\theta \wedge \Diamond\psi$ if it is not already a prime. The following lemmas are sufficient for our purpose.

Lemma 4.1 *Let* $\varphi_i : \sigma$, $1 \le i \le n$ *be prime assertions.* $\Box \bigvee_{1 \le i \le n} \varphi_i$ *is provable to be equivalent to a finite disjunction of prime assertions.*

Proof By induction on n.

$n = 1$:

$$\begin{aligned} \vdash \Box\varphi_1 &= \Box(\varphi_1 \vee \mathbf{f}) \\ &= \Box(\varphi_1 \vee \mathbf{f}) \wedge (\Diamond\varphi_1 \vee \Box\mathbf{f}) && (\Box) \\ &= [\Box(\varphi_1 \vee \mathbf{f}) \wedge \Diamond\varphi_1] \vee [\Box(\varphi_1 \vee \mathbf{f}) \wedge \Box\mathbf{f}] \\ &= \Box\varphi_1 \wedge \Diamond\varphi_1. \end{aligned}$$

Induction step:

$$\vdash \Box(\bigvee_{i=1}^{n} \varphi_i) = \Box(\bigvee_{i=1}^{n} \varphi_i) \wedge (\Diamond\varphi_1 \vee \Box(\bigvee_{2\le i\le n} \varphi_i)) \qquad (\Box)$$

$$= (\Diamond\varphi_1 \wedge \Box \bigvee_{i=1}^{n} \varphi_i) \vee \Box(\bigvee_{2\le i\le n} \varphi_i)$$

$$= (\Diamond\varphi_1 \wedge \Diamond\varphi_2 \wedge \Box \bigvee_{i=1}^{n} \varphi_i) \vee \Box(\bigvee_{i\neq 1} \varphi_i) \vee \Box(\bigvee_{i\neq 2} \varphi_i)$$

$$\cdots$$

$$= (\Box \bigvee_{i=1}^{n} \varphi_i \wedge \bigwedge_{1\le i\le n} \Diamond\varphi_i) \vee \bigvee_{1\le i\le n} (\Box \bigvee_{j\neq i} \varphi_j).$$

By the induction hypothesis each $\Box \bigvee_{j\neq i} \varphi_j$ has a prime normal form, hence so does $\Box \bigvee_{i=1}^{n} \varphi_i$.

□

We can derive from the above proof the following formula.

$$(\mathbf{P} - \Box) \qquad \Box \bigvee_{i\in I} \varphi_i = \bigvee_{J\subseteq I} (\Box \bigvee_{j\in J} \varphi_j \wedge \bigwedge_{j\in J} \Diamond\varphi_j)$$

Using the same method we can prove

Lemma 4.2 *Let $\varphi_i : \sigma$, $1 \le i \le n$ be prime assertions. Then*

$$\Box \bigvee_{1\le i\le n} \varphi_i \wedge \bigwedge_{j\in J} \Diamond\varphi_j$$

is provable to be equivalent to a finite disjunction of prime assertions, where

$$J \subseteq \{1, 2, \cdots, n\}.$$

Lemma 4.3 *Every assertion of the form*

$$\Box(\bigvee_{i=1}^{n} \varphi_i) \wedge \bigwedge_{i=1}^{n} \Diamond\varphi_i$$

is equivalent to a finite disjunction of prime assertions, where φ_i's are not necessarily prime.

Proof By assumption, for each i, there are ψ_{ij}, prime assertions such that

$$\vdash \varphi_i = \bigvee_{1\leq j\leq n_i} \psi_{i,j}.$$

We have

$$\begin{aligned}
\vdash \Box(\bigvee_{i=1}^{n} \varphi_i) \wedge \bigwedge_{i=1}^{n} \Diamond\varphi_i &= \Box(\bigvee_{i=1}^{n}\bigvee_{j=1}^{n_i} \psi_{i,j}) \wedge \bigwedge_{i=1}^{n} \Diamond(\bigvee_{j=1}^{n_i} \psi_{i,j}) \\
&= \Box(\bigvee_{i=1}^{n}\bigvee_{j=1}^{n_i} \psi_{i,j}) \wedge \bigwedge_{i=1}^{n}\bigvee_{j=1}^{n_i} \Diamond\psi_{i,j} \\
&= \bigvee_{f\in K} (\Box(\bigvee_{i=1}^{n}\bigvee_{j=1}^{n_i} \psi_{i,j}) \wedge \bigwedge_{i=1}^{n} \Diamond\psi_{i,f(i)}),
\end{aligned}$$

where $K = \{f : \overline{n} \to \overline{\max\{n_i \mid 1 \leq i \leq n\}} \mid \forall i.\ f(i) \in \overline{n_i}\}$.

From Lemma 4.2 we know that each term in $\bigvee_{f\in K}$ is equivalent to a finite disjunction of prime assertions, which concludes the proof of Lemma 4.3.

□

Lemma 4.4 *For any prime assertion*

$$\Box(\bigvee_{i=1}^{n} \varphi_i) \wedge \bigwedge_{i=1}^{n} \Diamond\varphi_i,$$

$$\left[\Box(\bigvee_{i=1}^{n} \varphi_i) \wedge \bigwedge_{i=1}^{n} \Diamond\varphi_i\right] \wedge \Diamond\psi$$

is equivalent to a finite disjunction of prime assertions.

Proof By ($\Diamond$) we see that

$$\begin{aligned}
&(\Box(\bigvee_{i=1}^{n} \varphi_i) \wedge \bigwedge_{i=1}^{n} \Diamond\varphi_i) \wedge \Diamond\psi \\
&= (\Box(\bigvee_{i=1}^{n} \varphi_i) \wedge \bigwedge_{i=1}^{n} \Diamond\varphi_i) \wedge \Diamond(\psi \wedge (\bigvee_{i=1}^{n} \varphi_i)) \\
&= \bigvee_{j=1}^{n} \Box((\bigvee_{i=1}^{n} \varphi_i) \vee (\varphi_j \wedge \psi)) \wedge \bigwedge_{i=1}^{n} \Diamond\varphi_i \wedge \Diamond(\psi \wedge \varphi_j).
\end{aligned}$$

By Lemma 4.3 the above assertion can be equivalently transformed into a finite disjunction of prime assertions.

□

Lemma 4.5 *For any prime assertion* $\Box(\bigvee_{i=1}^{n}\varphi_i)\wedge\bigwedge_{i=1}^{n}\Diamond\varphi_i$,

$$(\Box(\bigvee_{i=1}^{n}\varphi_i)\wedge\bigwedge_{i=1}^{n}\Diamond\varphi_i)\wedge\Box\psi$$

is equivalent to a finite disjunction of prime assertions.

Proof

$$\begin{aligned}\vdash(\Box(\bigvee_{i=1}^{n}\varphi_i)\wedge\bigwedge_{i=1}^{n}\Diamond\varphi_i)\wedge\Box\psi &= \Box((\bigvee_{i=1}^{n}\varphi_i)\wedge\psi)\wedge\bigwedge_{i=1}^{n}(\Diamond\varphi_i\wedge\Box\psi)\\ &= \Box\bigvee_{i=1}^{n}(\varphi_i\wedge\psi)\wedge\bigwedge_{i=1}^{n}\Diamond(\varphi_i\wedge\psi).\end{aligned}$$

By Lemma 4.4, the above assertion can be equivalently transformed into a finite disjunction of prime assertions.

□

Now we have

Proposition 4.5 *Plotkin powerdomain preserves property* p_0, p_1, *and* p_2.

Proof Property (p_1) follows from Lemma 4.1 to 4.5. The proof of (p_0) is given below.

$$\begin{aligned}&[\![(\Box\bigvee_{1\le i\le n}\varphi_i)\wedge\bigwedge_{1\le i\le n}\Diamond\varphi_i]\!]\subseteq[\![(\Box\bigvee_{1\le j\le m}\psi_j)\wedge\bigwedge_{1\le j\le m}\Diamond\psi_j]\!]\\
&\Longrightarrow[\![(\Box\bigvee_{1\le i\le n}\varphi_i)\wedge\bigwedge_{1\le i\le n}\Diamond\varphi_i]\!]\subseteq[\![\Box\bigvee_{1\le j\le m}\psi_j]\!]\cap\bigcap_{1\le j\le m}[\![\Diamond\psi_j]\!]\\
&\Longrightarrow\{\check{\psi}_j\mid 1\le j\le m\}\sqsubseteq_0\{\check{\varphi}_i\mid 1\le i\le n\}\ \&\\
&\qquad\{\check{\psi}_j\mid 1\le j\le m\}\sqsubseteq_1\{\check{\varphi}_i\mid 1\le i\le n\}\\
&\Longrightarrow\forall i\,\exists j.\ \vdash\varphi_i\le\psi_j\ \ \&\ \forall j\,\exists i.\ \vdash\varphi_i\le\psi_j\\
&\Longrightarrow\vdash\Box\bigvee_{1\le i\le n}\varphi_i\le\Box\bigvee_{1\le j\le m}\psi_j\ \ \&\ \vdash\bigwedge_{1\le i\le n}\Diamond\varphi_i\le\bigwedge_{1\le j\le m}\Diamond\psi_j\\
&\Longrightarrow\vdash\Box\bigvee_{1\le i\le n}\varphi_i\wedge\bigwedge_{1\le i\le n}\Diamond\varphi_i\le\Box\bigvee_{1\le j\le m}\psi_j\wedge\bigwedge_{1\le j\le m}\Diamond\psi_j\end{aligned}$$

□

The completeness of the proof system now follows from Proposition 4.1 to Proposition 4.7 by a structural induction on type structures, with each proposition taking care of one case. So we have

Theorem 4.2 *The proof system is complete.*

For expressiveness of our assertion language we have

Theorem 4.3 *For each σ,*

$$[\![\ ]\!]_\sigma : (\mathcal{A}_\sigma/=,\ \leq_\sigma) \to (\mathbf{K}\Omega\mathcal{D}(\sigma),\ \subseteq)$$

is an isomorphism, where $\mathbf{K}\Omega(D)$ is the set of compact open sets of D.

Proof By the Completeness Theorem it is enough to show that

$$\forall P \in \mathbf{P}\Omega(\mathcal{D}(\sigma))\ \exists \varphi \in \mathcal{A}_\sigma.\ \mathbf{P}(\varphi)\ \&\ [\![\varphi]\!]_\sigma = P,$$

where $\mathbf{P}\Omega(D)$ is the set of complete primes of the lattice $\Omega(D)$. This can be done by a structural induction.

□

It is time we had some examples. In Abramsky's framework for Scott domains [Ab87], the conjunction of prime assertions is again prime. Now we present an example to show that, in our framework, the conjunction of prime assertions need not be equivalent to a prime assertion. Consider the type $\mathcal{P}_P((\mathbf{1}_\perp + \mathbf{1}_\perp) \times (\mathbf{1}_\perp + \mathbf{1}_\perp))$. Write

$$p \equiv \mathit{inl}\,\mathbf{t}_\perp : \mathbf{1}_\perp + \mathbf{1}_\perp,$$
$$q \equiv \mathit{inr}\,\mathbf{t}_\perp : \mathbf{1}_\perp + \mathbf{1}_\perp.$$

Obviously $\vdash p \wedge q \leq \mathbf{f}$ and

$$\Box(p \times \mathbf{t} \vee q \times \mathbf{t}) \wedge \Diamond p \times \mathbf{t} \wedge \Diamond q \times \mathbf{t},$$
$$\Box(\mathbf{t} \times p \vee \mathbf{t} \times q) \wedge \Diamond \mathbf{t} \times p \wedge \Diamond \mathbf{t} \times q$$

are prime assertions.

We have

$$\Box(p\times\mathbf{t}\vee q\times\mathbf{t})\wedge\Diamond p\times\mathbf{t}\wedge\Diamond q\times\mathbf{t}$$

$$\wedge\Box(\mathbf{t}\times p\vee\mathbf{t}\times q)\wedge\Diamond\mathbf{t}\times p\wedge\Diamond\mathbf{t}\times q$$

$$=\Box(p\times p\vee p\times q\vee q\times p\vee q\times q)\wedge\Diamond p\times\mathbf{t}\wedge\Diamond q\times\mathbf{t}\wedge\Diamond\mathbf{t}\times p\wedge\Diamond\mathbf{t}\times q$$

$$=(\textstyle\bigvee_i\alpha_i)\wedge\Diamond p\times\mathbf{t}\wedge\Diamond q\times\mathbf{t}\wedge\Diamond\mathbf{t}\times p\wedge\Diamond\mathbf{t}\times q$$

by formula $(\mathbf{P}-\Box)$, given after Lemma 4.1 α_i's are prime

$$=\Box(p\times q\vee q\times p)\wedge\Diamond p\times q\wedge\Diamond q\times p$$

$$\vee\ \Box(p\times p\vee q\times q)\wedge\Diamond p\times p\wedge\Diamond q\times q$$

$$\vee\ \Box(p\times q\vee q\times p\vee p\times p)\wedge\Diamond p\times q\wedge\Diamond q\times p\wedge\Diamond p\times p$$

$$\vee\ \Box(p\times q\vee q\times p\vee q\times q)\wedge\Diamond p\times q\wedge\Diamond q\times p\wedge\Diamond q\times q$$

$$\vee\ \Box(p\times q\vee q\times q\vee p\times p)\wedge\Diamond p\times q\wedge\Diamond q\times q\wedge\Diamond p\times p$$

$$\vee\ \Box(q\times p\vee q\times q\vee p\times p)\wedge\Diamond q\times p\wedge\Diamond q\times q\wedge\Diamond p\times p$$

$$\vee\ \Box(q\times p\vee p\times q\vee q\times q\vee p\times p)\wedge\Diamond q\times p\wedge\Diamond p\times q$$

$$\wedge\Diamond q\times q\wedge\Diamond p\times p.$$

The following remarks explain the above derivation.

1. By formula $(\mathbf{P}-\Box)$ given after Lemma 4.1, all possible primes formed from subset of $\{\varphi_i\mid i\in I\}$ should be in the disjunction. We omitted some of them in the last formula. But the equivalence still holds, because we have, for example,

$$\vdash\Box(p\times p\vee p\times q)\wedge\Diamond q\times\mathbf{t}\leq\Diamond(p\times p\vee p\times q)\wedge q\times\mathbf{t}\leq\mathbf{f},$$

that is, all the omitted assertions are provable to be equivalent to $\mathbf{f}$ when put in conjunction with

$$\Diamond p\times\mathbf{t}\wedge\Diamond q\times\mathbf{t}\wedge\Diamond\mathbf{t}\times p\wedge\Diamond\mathbf{t}\times q.$$

2. For those α_i's which remain in the last derivation,

$$\vdash\alpha\wedge(\Diamond p\times\mathbf{t}\wedge\Diamond q\times\mathbf{t}\wedge\Diamond\mathbf{t}\times p\wedge\Diamond\mathbf{t}\times q)=\alpha$$

because of obvious reasons such as

$$\vdash\Diamond p\times q\wedge\Diamond p\times\mathbf{t}=\Diamond p\times q.$$

3. The prime assertions in the last formula are pairwise inconsistent. We have, for example,

$$\begin{aligned}
&\Box(p \times q \vee q \times q \vee p \times p) \wedge \Diamond p \times q \wedge \Diamond q \times q \wedge \Diamond p \times p \\
&\qquad \wedge \ \Box(q \times p \vee q \times q \vee p \times p) \wedge \Diamond q \times p \wedge \Diamond q \times q \wedge \Diamond p \times p \\
&\leq \Box(p \times q \vee q \times q \vee p \times p) \wedge \Diamond q \times p \\
&\leq \Diamond q \times p \ \wedge \ (p \times q \vee q \times q \vee p \times p) \\
&\leq \mathbf{f}.
\end{aligned}$$

4. We really need a machine to carry out this example!

4.5 Pure Relative-Completeness

The proof system presented in the previous sections is concerned with the family of closed types. In general, it is desirable to allow the possibility of variable types in a proof system. There are two ways to formulate proof systems with variable types.

The first approach is to treat type variables as 'holes' in the proof system, to be filled by other proof systems. This allows one to focus on some components of the proof system and leave the rest unspecified. To illustrate the idea by an example, consider the recursively defined type $rec\ t.((\mathbf{1}_\perp + t) + \alpha)$, where α is an unbound type variable (a varaible type). It is reasonable to have

$$\mathit{inl}\,(\,\mathit{inr}\,\mathbf{t}_\perp) \vee \mathit{inr}\,(\quad)_\alpha$$

as an assertion of this type. That is, assertions also have holes in them to be filled by those of the appropriate types. Formally, the effect of this is the same as introducing assertion holes

$$(\quad)_\alpha \ : \alpha$$

for a variable type α and leave the rest of the assertion formation rules unchanged. The key point here is that we do not reason about 'holes'; Instead, the holes are only there to be filled by other proof systems.

We call this kind of proof systems *proof systems with parameters*. A proof system with parameters is sound and pure relatively-complete if when we supplement the proof system with sound and complete proof systems for the variable types we get a sound and complete system. Here 'sound' and 'complete' are basically in the same sense as specified in the previous sections.

The following soundness and pure relative-completeness theorem is a rather easy corollary of the soundness and completeness results given in the previous section.

Theorem 4.4 *Proof systems with parameters are sound and pure relative-complete.*

It merits some explanation why the term 'pure relative-completeness' is adopted instead of relative completeness. By relative completeness one usually refers to the first completeness result discovered by Cook [Co78] for Hoare's proof system of the while-programs. The completeness of Hoare logic is achieved by assuming an oracle for deciding facts about integers. However, there was nothing to ensure that the facts about integers used were about integers in the programming language, and weren't somehow transferring themselves to other parts of the system. In fact, as pointed out by Apt [Ap81], even one restricts the assertions to only 'true' and 'false' (for which there is clearly a sound and complete proof system), Hoare logic can still be incomplete. This shows facts about integers were definitely used somewhere else in the proof system of Hoare logic.

The completeness of proof systems with parameters is a totally different story. If you substitute parameters by sound and complete proof systems, you will get a sound and complete proof system. Here, the type system has the effect of confining the oracle to reporting on what it is supposed to know about, and, moreover, the system does not care what the oracle is reporting about, since a parameter type can be substituted by any closed type expression. It is in this sense that proof systems with parameters are called 'pure relative-complete'.

The foregoing discussion does address some unresolved interesting issues which we leave as research topics. If we were to derive some kind of logic for the while-programs based on our framework, what do we have to change to get something close to Hoare logic? It is not clear at all at the moment, considering the different kind of relative completeness we get, as well as a fact pointed out by Robinson [Ro87] that the familiar denotational semantics for the while-programs does not offer a Scott domain from which Hoare logic can be derived.

Pure relative-completeness is not the only possible notion on which we can discuss completeness, however. There is the second, different, approach which we introduce now. Instead of introducing assertion holes, we introduce variable assertions for variable types:

$$x_0^\sigma,\ x_1^\sigma,\ \cdots : t$$

The rest of the assertion formation rules remain the same except that when we write $\varphi : \sigma$, both φ and σ may contain variables. Write $\mathcal{VA}_\sigma$ for the collection of assertions of a type σ.

Although the formulation of the axioms and rules is the same, it makes a broader sense as assertions can now have free variables in them. We have, for example, as an instance of the axiom

$$\mathit{inl}\,(\varphi \vee \psi) = (\mathit{inl}\varphi) \vee (\mathit{inl}\psi)$$

the formula

$$\mathit{inl}\,(x \vee y) = (\mathit{inl}x) \vee (\mathit{inl}y).$$

To give a semantic interpretation of assertions with variables, we introduce environments, which are maps from type variables and typed variable-assertions to SFP domains and their compact open sets.

Definition 4.4 *A type environment θ is a function from type variables to SFP domains. Given a type environment, θ, an assertion environment ρ is a function*

$$\rho : \bigcup_\sigma \{ x^\sigma \mid x^\sigma : \sigma \} \to \bigcup_\sigma \{ K \mid K \in \mathbf{K\Omega}\mathcal{D}(\sigma) \}$$

such that $\forall x^\sigma.\ \rho(x^\sigma) \in \mathbf{K\Omega}\mathcal{D}(\sigma)$, and the domain $\mathcal{D}(\sigma)$ is determined by θ. Write Env for the set of environment pairs (θ, ρ), with θ a type environment and ρ an assertion environment.

The interpretation of assertions is given by taking into account the environments. For each type expression σ (which may contain type variables) we define an *interpretation function*

$$[\![\ \ ]\!]_\sigma : \mathcal{VA}_\sigma \to [\mathit{Env} \to \mathbf{K\Omega}(\mathcal{D}(\sigma))]$$

in a structural way. For example, for each type σ, we have

$$\begin{aligned}
&[\![\mathbf{t}]\!]_\sigma\varepsilon = \mathcal{D}(\sigma),\\
&[\![\mathbf{f}]\!]_\sigma\varepsilon = \emptyset,\\
&[\![\varphi \vee \psi]\!]_\sigma\varepsilon = [\![\varphi]\!]_\sigma\varepsilon \cup [\![\psi]\!]_\sigma\varepsilon,\\
&[\![\varphi \wedge \psi]\!]_\sigma\varepsilon = [\![\varphi]\!]_\sigma\varepsilon \cap [\![\psi]\!]_\sigma\varepsilon,\\
&[\![x^\sigma]\!]\varepsilon = \varepsilon(x^\sigma).
\end{aligned}$$

A similar definition is given for type constructions. Under this interpretation, the definition of soundness and completeness needs to be adjusted. We say that the proof system is *sound* if for assertions φ, ψ of any type σ (with variables), $\vdash \varphi \leq \psi$ implies $[\![\varphi]\!]\varepsilon \subseteq [\![\psi]\!]\varepsilon$ for any environment ε. We say that the proof system is *complete* if given that $[\![\varphi]\!]\varepsilon \subseteq [\![\psi]\!]\varepsilon$ holds for any environment ε, we have $\vdash \varphi \leq \psi$.

It was a bit of a surprise to me that completeness under this formulation can not be settled straight away without some change of the proof system. Although it is very tempting to achieve such a completeness, I will not offer a reformulation and a proof of the completeness here (If I try to get everything settled, I may never be able to finish the book). However, examples are given below showing why the proof system as it stands is not complete for variable assertions.

Consider the type $t_\perp + t_\perp$, where t is a type variable. Then with x, $y : t$,

$$\mathit{inl}\, x_\perp \wedge \mathit{inr}\, y_\perp \leq \mathbf{f}$$

is a formula valid under all environments. However, it cannot be derived from the proof system. This shows some generalization of the predicate $\mathbf{Q}$ is necessary.

Note also that under a proper type assignment,

$$(\mathit{inl}\, x \wedge \mathit{inr}\, x) \to u \vee v = (\mathit{inl}\, x \wedge \mathit{inr}\, x) \to u \;\vee\; (\mathit{inl}\, x \wedge \mathit{inr}\, x) \to v$$

and

$$(x \to \mathbf{f}) \to u \vee v = (x \to \mathbf{f}) \to u \;\vee\; (x \to \mathbf{f}) \to v$$

are both valid under all environments since $\mathit{inl}\, x \wedge \mathit{inr}\, x$ and $x \to \mathbf{f}$ are equivalent to either $\mathbf{t}$, or $\mathbf{f}$, depending on the environment. The above formulae can not be derived either. That shows the necessity to reformulate the predicate $\mathbf{P}$.

4.6 An Application

As an application, we show how our framework can generate the assertions in Brookes' proof system [Br85] but without using labels.

The syntax of the parallel programming language is specified by

$$\Gamma ::= \textbf{skip} \mid I := E \mid \Gamma_1; \Gamma_2 \mid \Gamma_1 \parallel \Gamma_2 \mid \textbf{if } B \textbf{ then } \Gamma_1 \textbf{ else } \Gamma_2 \mid \textbf{while } B \textbf{ do } \Gamma.$$

Here to avoid a problem with completeness as caused by the command

$$\textbf{await } \alpha : x > 0 \textbf{ then } \beta : (\textbf{while } \gamma : x > 0 \textbf{ do } \delta : x := x - 1),$$

we simply keep away from the await statements at this point.

We use an unlabelled transition system

$$((\textbf{Com} \times \Sigma) \cup \Sigma, \rightarrow)$$

to give an operational semantics for the programming language. An element of $\textbf{Com} \times \Sigma$, say, $\langle \Gamma, \sigma \rangle$, represents a stage (configuration) in a computation at which the remaining command to be executed is Γ, and the current state is σ. Configurations of the form $\langle \textbf{null}, \sigma \rangle$ used by Brookes are denoted by a termination state σ. The relation $\rightarrow$ is specified by the following rules:

Atomic Transitions

$$(S1) \quad \langle \textbf{skip}, \sigma \rangle \rightarrow \sigma$$

$$(S2) \quad \langle I := E, \sigma \rangle \rightarrow \sigma[E/I]$$

$$(S3) \quad \frac{\langle \Gamma_1, \sigma \rangle \rightarrow \langle \Gamma_1', \sigma' \rangle}{\langle \Gamma_1; \Gamma_2, \sigma \rangle \rightarrow \langle \Gamma_1'; \Gamma_2, \sigma' \rangle}$$

$$(S4) \quad \frac{\langle \Gamma_1, \sigma \rangle \rightarrow \langle \Gamma_1', \sigma' \rangle \quad \sigma \models B}{\langle \textbf{if } B \textbf{ then } \Gamma_1 \textbf{ else } \Gamma_2, \sigma \rangle \rightarrow \langle \Gamma_1', \sigma' \rangle}$$

$$(S5) \quad \frac{\langle \Gamma_2, \sigma \rangle \rightarrow \langle \Gamma_2', \sigma' \rangle \quad \sigma \models \neg B}{\langle \textbf{if } B \textbf{ then } \Gamma_1 \textbf{ else } \Gamma_2, \sigma \rangle \rightarrow \langle \Gamma_2', \sigma' \rangle}$$

$$(S6) \quad \frac{\langle \Gamma, \sigma \rangle \rightarrow \langle \Gamma', \sigma' \rangle \quad \sigma \models B}{\langle \textbf{while } B \textbf{ do } \Gamma, \sigma \rangle \rightarrow \langle \Gamma'; \textbf{while } B \textbf{ do } \Gamma, \sigma' \rangle}$$

$$(S7) \quad \frac{\sigma \models \neg B}{\langle \textbf{while } B \textbf{ do } \Gamma, \sigma \rangle \rightarrow \sigma}$$

$$(S8) \quad \frac{\langle \Gamma_1, \sigma \rangle \rightarrow \langle \Gamma_1', \sigma' \rangle}{\langle \Gamma_1 \parallel \Gamma_2, \sigma \rangle \rightarrow \langle \Gamma_1' \parallel \Gamma_2, \sigma' \rangle}$$

$$(S9) \quad \frac{\langle \Gamma_2, \sigma \rangle \rightarrow \langle \Gamma_2', \sigma' \rangle}{\langle \Gamma_1 \parallel \Gamma_2, \sigma \rangle \rightarrow \langle \Gamma_1 \parallel \Gamma_2', \sigma' \rangle}$$

When $\langle\Gamma, \sigma\rangle \rightarrow \langle\Gamma', \sigma'\rangle$ appears above the line in a transition rule it is understood that $\langle\Gamma', \sigma'\rangle$ may take the form σ'. For example, the transition system has a rule instance

$$\frac{\langle\Gamma_1, \sigma\rangle \rightarrow \sigma'}{\langle\Gamma_1; \Gamma_2, \sigma\rangle \rightarrow \langle\Gamma_2, \sigma'\rangle}.$$

We now show that the assertions Brookes used (but without labels) can be derived from the domain for the denotational semantics of the parallel programming language, that is, the domain of resumptions

$$R \cong S \rightarrow \mathcal{P}_P[(S \times R) + S]$$

proposed by Plotkin [Pl76]. Here S is the domain of states. It is treated as a type variable since this leaves some option as to what should be the type of states. Therefore, the domain of resumptions is specified by the type the type

$$r = rec\, t.(\alpha \rightarrow \mathcal{P}_P[(\alpha \times t) + \alpha]).$$

For convenience let P and P_i's be assertions for states. According to the assertion formation rules, assertions of type $(\alpha \times r) + \alpha$ are given by

$$u ::= inr(P) \mid \; inl(P \times \varphi) \mid \; u \wedge u \mid \; u \vee u \mid \; \mathbf{t} \;\mid \mathbf{f},$$

assertions of type $\mathcal{P}_P[(\alpha \times r) + \alpha]$ by

$$w ::= \Box u \mid \Diamond u \mid w \wedge w \mid w \vee w \mid \; \mathbf{t} \;\mid \mathbf{f},$$

and, finally, assertions of type r are given by

$$\varphi ::= P \rightarrow w \mid \varphi_1 \wedge \varphi_2 \mid \varphi_1 \vee \varphi_2 \mid \; \mathbf{t} \;\mid \mathbf{f}.$$

Now we can see that assertions of the following form

$$P \rightarrow \Box(\bigvee_{i=1}^{n} inl\,(P_i \times \varphi_i)) \wedge \bigwedge_{i=1}^{n} \Diamond inl\,(P_i \times \varphi_i)$$

play the same role as assertions

$$P \Sigma_{i=1}^{n} \alpha_i P_i \varphi_i$$

did in Brookes' proof system.

To see this, keep in mind that an assertion represents an open set of a domain, and the denotation of a program is an element of the domain. Thus our framework carries with it a natural definition that program C satisfies assertion A if $[\![C]\!] \in [\![A]\!]$. For the programming language under consideration, a command Γ satisfies

$$P \rightarrow \Box(\bigvee_{i=1}^{n} \mathit{inl}(P_i \times \varphi_i)) \wedge \bigwedge_{i=1}^{n} \Diamond \mathit{inl}(P_i \times \varphi_i),$$

written

$$\Gamma \models P \rightarrow \Box(\bigvee_{i=1}^{n} \mathit{inl}(P_i \times \varphi_i)) \wedge \bigwedge_{i=1}^{n} \Diamond \mathit{inl}(P_i \times \varphi_i),$$

if

$$[\![\Gamma]\!] \in [\![P \rightarrow \Box(\bigvee_{i=1}^{n} \mathit{inl}\,(P_i \times \varphi_i)) \wedge \bigwedge_{i=1}^{n} \Diamond \mathit{inl}\,(P_i \times \varphi_i)]\!].$$

However this is the same as saying, in terms of operational semantics (here we have assumed the equivalence of the denotational semantics and the operational semantics), that we have the following two statements. First,

$$\forall \sigma. \left[(\sigma \models P) \Longrightarrow \{\langle \Gamma', \sigma' \rangle \mid \langle \Gamma, \sigma \rangle \rightarrow \langle \Gamma', \sigma' \rangle\} \models \Box(\bigvee_{i=1}^{n} \mathit{inl}\,(P_i \times \varphi_i)) \right],$$

in other words,

$$\forall \sigma. \left[(\sigma \models P) \& \langle \Gamma, \sigma \rangle \rightarrow \langle \Gamma', \sigma' \rangle \Longrightarrow \langle \Gamma', \sigma' \rangle \models (\bigvee_{i=1}^{n} \mathit{inl}\,(P_i \times \varphi_i)) \right].$$

And second,

$$\forall \sigma. \left[(\sigma \models P) \Longrightarrow \{\langle \Gamma', \sigma' \rangle \mid \langle \Gamma, \sigma \rangle \rightarrow \langle \Gamma', \sigma' \rangle\} \models \bigwedge_{i=1}^{n} \Diamond \mathit{inl}\,(P_i \times \varphi_i) \right],$$

or

$$\forall i\, \forall \sigma. [(\sigma \models P) \Longrightarrow \{\langle \Gamma', \sigma' \rangle \mid \langle \Gamma, \sigma \rangle \rightarrow \langle \Gamma', \sigma' \rangle\} \models \Diamond \mathit{inl}\,(P_i \times \varphi_i)].$$

In summary,

$$\Gamma \models P \rightarrow \Box(\bigvee_{i=1}^{n} \mathit{inl}\,(P_i \times \varphi_i)) \wedge \bigwedge_{i=1}^{n} \Diamond \mathit{inl}\,(P_i \times \varphi_i)$$

if

1. $\forall \sigma.\ (\sigma \models P)\ \&\ \langle \Gamma, \sigma \rangle \rightarrow \langle \Gamma', \sigma' \rangle$ implies
$$\exists i.\ (1 \leq i \leq n)\ \&\ \sigma' \models P_i\ \&\ \Gamma' \models \varphi_i, \quad \text{and}$$
2. $\forall i \forall \sigma.\ (\sigma \models P)$ implies
$$\exists \langle \Gamma', \sigma' \rangle.\ \langle \Gamma, \sigma \rangle \rightarrow \langle \Gamma', \sigma' \rangle\ \&\ \Gamma' \models \varphi_i\ \&\ \sigma' \models P_i.$$

This is exactly a restatement of the definition for

$$\models \Gamma\ \mathbf{sat}\ \ P\Sigma_{i=1}^{n} \alpha_i P_i \varphi_i$$

in [Br85] without using labels.

4.7 Non-entailment

Sometimes it is necessary to be able to reason about negative information. We actually have already seen such a need in predicate **Q**. It captures some assertions which are *not* equivalent to **t**. Another use of negative information is in predicate **P**. An assertion φ being prime, written $\mathbf{P}(\varphi)$, implies in particular that φ is not equivalent to **f**.

The purpose of this section is to give an axiomatization of the non-entailment $\not\leq$ between assertions. A sound and complete proof system for non-entailment is provided. As a consequence we know that $\leq$ is decidable.

Because a lot of effort has already been spent in the previous sections on $\leq$, this section is going to be sketchy. Based on the results of the previous sections, it is a relatively easy step to prove the soundness and completeness of the proof system for non-entailment with closed types. The proof system is sound in the sense that whenever we can derive a formula $\varphi \not\leq \psi$ from the system, it must be the case that $[\![\varphi]\!] \not\subseteq [\![\psi]\!]$, under the same semantic interpretation for assertions given before. It is complete in the sense that if $[\![\varphi]\!] \not\subseteq [\![\psi]\!]$, then we must be able to derive the formula $\varphi \not\leq \psi$ from the system.

The proof system for $\not\leq$ is built on top of the type expressions, assertions, and proof rules given before. In the following a formula like $\varphi \not\leq \psi$ should be read as 'φ does not entail ψ.'

$$(\mathbf{t},\mathbf{f}) \qquad \mathbf{t} \not\leq \mathbf{f}$$

$$(\neg - \wedge) \qquad \frac{\varphi \not\leq \psi}{\varphi \not\leq \psi \wedge \psi'}$$

$$(\vee - \neg) \qquad \frac{\varphi \not\leq \psi}{\varphi \vee \varphi' \not\leq \psi}$$

$$(\wedge - \neg) \qquad \frac{\varphi \wedge \psi \not\leq \psi'}{\varphi \not\leq \psi'} \qquad\qquad \frac{\varphi \wedge \psi \not\leq \psi'}{\psi \not\leq \psi'}$$

$$(\neg - \vee) \qquad \frac{\varphi \not\leq \psi_1 \vee \psi_2}{\varphi \not\leq \psi_1} \qquad\qquad \frac{\varphi \not\leq \psi_1 \vee \psi_2}{\varphi \not\leq \psi_2}$$

$$(= - \neg) \qquad \frac{\varphi = \varphi' \qquad \varphi' \not\leq \psi' \qquad \psi' = \psi}{\varphi \not\leq \psi}$$

$$(\neg - \mathbf{P}) \qquad \frac{\mathbf{P}(\varphi) \qquad \forall i \in I.\, \varphi \not\leq \psi_i}{\varphi \not\leq \bigvee_{i \in I} \psi_i}$$

With respect to type constructions we have

$$(\neg + \mathbf{f}) \qquad \frac{\mathbf{t} \not\leq \varphi \qquad \mathbf{t} \not\leq \psi}{\mathit{inl}\,\varphi \wedge \mathit{inr}\,\psi \leq \mathbf{f}}$$

$$(\neg - +) \qquad \frac{\varphi \not\leq \psi}{\mathit{inl}\,\varphi \not\leq \mathit{inl}\,\psi} \qquad\qquad \frac{\varphi \not\leq \psi}{\mathit{inr}\,\varphi \not\leq \mathit{inr}\,\psi}$$

$$(\bot - \mathbf{t}) \qquad \mathbf{t} \not\leq \varphi_\bot$$

$$(\neg - \bot) \qquad \frac{\varphi \not\leq \psi}{\varphi_\bot \not\leq \psi_\bot}$$

$$(\neg - \Box) \qquad \frac{\varphi \not\leq \psi}{\Box\varphi \not\leq \Box\psi}$$

$$(\neg - \Diamond) \qquad \frac{\varphi \not\leq \psi}{\Diamond\varphi \not\leq \Diamond\psi}$$

In the following table assertions are all assumed to be prime, including their component assertions. Without this assumption some of the rules may not be valid.

$$(\neg - \times) \qquad \frac{\varphi \not\leq \varphi'}{\varphi \times \psi \not\leq \varphi' \times \psi'} \qquad\qquad \frac{\psi \not\leq \psi'}{\varphi \times \psi \not\leq \varphi' \times \psi'}$$

$$(\neg - \rightarrow) \qquad \frac{\bigwedge_{\varphi_i \leq \alpha} \psi_i \not\leq \beta}{\bigwedge_{i \in I} \varphi_i \rightarrow \psi_i \not\leq \alpha \rightarrow \beta}$$

$$(\neg\Diamond\Box) \qquad \frac{\bigwedge_{i \in I} \Diamond\varphi_i \not\leq \bigwedge_{j \in J} \Diamond\psi_j}{(\Box \bigvee_{i \in I} \varphi_i) \wedge \bigwedge_{i \in I} \Diamond\varphi_i \not\leq (\Box \bigvee_{j \in J} \psi_j) \wedge \bigwedge_{j \in J} \Diamond\psi_j}$$

$$\frac{\Box \bigvee_{i \in I} \varphi_i \not\leq \Box \bigvee_{j \in J} \psi_j}{(\Box \bigvee_{i \in I} \varphi_i) \wedge \bigwedge_{i \in I} \Diamond\varphi_i \not\leq (\Box \bigvee_{j \in J} \psi_j) \wedge \bigwedge_{j \in J} \Diamond\psi_j}$$

Chapter 5

A Mu-Calculus

The logical framework presented in the previous chapter is an axiomatization of the entailment on assertions using terms, types, and meta-predicates. These assertions were interpreted as compact open sets of the domains associated with the types. The expressive power of the logic is, therefore, not as strong as one sometimes would like. There is no assertion, for example, of the type of natural numbers to express 'the set of even numbers'. To make the logic more expressive some extra structure on assertions is needed. One of the possible ways to get a richer assertion language is to introduce fixed-point operators.

Fixed-points are useful because, as pointed out by Larsen [La90], greatest fixed-points can express safety properties and least fixed-points liveness properties of concurrent processes. For example, the least fixed-point of the function

$$\lambda Y.\, [a]\, \mathbf{f} \vee [a]\, Y,$$

written as $\mu y.\, [a]\, \mathbf{f} \vee [a]\, y$, expresses the property of an eventual deadlock. On the other hand, the greatest fixed-point, $\nu x.\, {<a>}\, \mathbf{t} \wedge [a]\, x$, of the function $\lambda X.\, {<a>}\, \mathbf{t} \wedge [a]\, X$ expresses the safety property 'always being able to perform the action a'. The reason is really simple since by unwinding the expressions we get the following semantic equalities:

$$\begin{aligned} &\mu y.\, [a]\, \mathbf{f} \vee [a]\, y \\ &= [a]\, \mathbf{f} \vee [a]\, [a]\, \mathbf{f} \vee [a]\, [a]\, [a]\, \mathbf{f} \vee \cdots, \end{aligned}$$

and

$$\nu x.\, <a>\mathbf{t} \wedge [a]\, x$$
$$=<a>\mathbf{t} \wedge [a] <a> \mathbf{t} \wedge [a]\,[a] <a> \mathbf{t} \wedge \cdots.$$

There has been quite a lot of work on reasoning about fixed-points, especially the least fixed-points, under various context. Typical among them are [Pr81], [Ko83], [Ni84], and [La90] in areas of dynamic logic, equational logic and Hennessy-Milner logic. However, different from the various systems of mu-calculi studied earlier, the rules for fixed-point induction (Park's induction rules) alone are not sufficient for the completeness of the mu-calculus we consider. This is because the type structure of the domain logic gives each assertion only one 'model', the open set it denotes. This unique character introduces some difficulty in achieving completeness since we cannot use the richness of the model to get hold of the possible structure a valid formula must assume. In this chapter we show that by introducing two more rules, the mu-calculus for the domain of integers is complete. We also discuss the expressiveness of the mu-calculus.

5.1 Recursive Assertions

To focus our attention on the least fixed-points of assertions we work on a language of type expressions without the powerdomains (they can be introduced without much change of the framework):

$$\sigma ::= \mathbf{1} \mid \sigma \times \tau \mid \sigma \rightarrow \tau \mid \sigma + \tau \mid \sigma_\perp \mid t \mid rec\, t.\sigma$$

where t is a type variable and σ, τ ranges over type expressions. Clearly each type σ can be interpreted as a Scott domain $\mathcal{D}(\sigma)$, as specified in the previous chapter.

Assertions of *closed* types are specified by the following table. Here when writing $\varphi(x)$, we mean that 'x is *a possible* variable in φ'. Each type is associated with a countable number of free variables and these variables are assertions. If $\varphi(x^\sigma)$ is an assertion of a type σ and x^σ is a free variable, we can form $\mu x^\sigma.\, \varphi(x^\sigma)$, to express the least fixed-point of $\varphi(x^\sigma)$. Assertions of this kind are called μ-assertions. For a μ-assertion of the form $\mu x.\, p \vee R(x)$ with p closed, we informally call $R(x)$ a *recursive unit* and p a *contributor.*

For each closed type σ we have

$$(\mathbf{t},\mathbf{f}) \qquad \mathbf{t},\mathbf{f} : \sigma$$

$$(Var) \qquad x_0^\sigma,\ x_1^\sigma,\ \cdots : \sigma$$

$$(\wedge - \vee) \qquad \frac{\varphi,\ \psi : \sigma}{\varphi \wedge \psi : \sigma \qquad \varphi \vee \psi : \sigma}$$

$$(\mu) \qquad \frac{\varphi(x^\sigma) : \sigma}{\mu x^\sigma.\ \varphi(x^\sigma) : \sigma}$$

$$(A) \qquad \frac{\varphi \in \mathcal{A}_\sigma}{\varphi : \sigma}$$

When no confusion arises, a variable x^σ is written simply as x.

With respect to type constructions the assertions are introduced by the following rules.

$$(\times) \qquad \frac{\varphi : \sigma \qquad \psi : \tau}{\varphi \times \psi : \sigma \times \tau}$$

$$(\rightarrow) \qquad \frac{\varphi \in \mathcal{A}_\sigma \qquad \psi : \tau}{\varphi \rightarrow \psi : \sigma \rightarrow \tau}$$

$$(+) \qquad \frac{\varphi : \sigma \qquad \psi : \tau}{\mathit{inl}\ \varphi : \sigma + \tau \qquad \mathit{inr}\ \psi : \sigma + \tau}$$

$$(\perp) \qquad \frac{\varphi : \sigma}{\varphi_\perp : \sigma_\perp}$$

$$(rec) \qquad \frac{\varphi : \sigma[rec\, t.\sigma/t]}{\varphi : rec\, t.\sigma}$$

Write $\mathcal{M}_\sigma$ for the set $\{\varphi \mid \varphi : \sigma\}$.

Example 5.1

$$\mu x.\ [\mathit{inl}(\mathbf{t}_\perp) \vee \mathit{inr}\ (\mathit{inr}\ (x))]$$

is a μ-assertion of type rec $t.\ (\mathbf{1}_\perp + t)$ which denotes the set of even numbers.

Note that the assertions are generated in the same manner as those for the propositional version in Chapter4 except for $(\rightarrow)$, where φ is required to be in $\mathcal{A}_\sigma$. There are two reasons for this restriction. First, if free variables are allowed immediately on the left of $\rightarrow$, it can cause the interpretation function non-monotonic (*e.g.* $x \rightarrow \mathbf{f}$), as is necessary for the existence of least fixed-points. Secondly, note the interpretation of a μ-assertion need not be a compact open set any more. Hence if the assertion φ in $\varphi \rightarrow \psi$

is allowed to be a μ-assertion, it can lead to $\varphi \to \psi$ being non-open, a situation we want to avoid at this stage.

Example 5.2 *The following assertion shows if we did not require $\varphi \in \mathcal{A}_\sigma$ in ($\to$) we would have allowed an assertion such as*

$$(\,\mu x.\,[\,\mathit{inl}(\mathbf{t}_\perp) \vee \mathit{inr}\,(\,x\,)\,]\,) \to \mathbf{t}_\perp$$

of type $(\mathit{rec}\, t.\,(\,\mathbf{1}_\perp + t\,)) \to \mathbf{1}_\perp$, which would, according to the the interpretation given later, represent the set

$$\{\, f : \mathcal{N}_\perp \to \mathcal{O} \mid \omega \subseteq f^{-1}(\top)\,\}.$$

This set is not open since a limit (of a chain of functions) being in this set does not imply that a finite approximation of the limit is also in the set.

To give a semantic interpretation of the μ-assertions we introduce *environments,* which are maps from type variables to compact open sets.

Definition 5.1 *An environment ρ is a function*

$$\rho\ :\ \bigcup_\sigma \{\, x^\sigma \mid x^\sigma : \sigma\,\} \to \bigcup_\sigma \{\, K \mid K \in \mathbf{K}\Omega\mathcal{D}\,(\,\sigma\,)\,\}$$

such that $\forall x^\sigma.\ \rho(\,x^\sigma\,) \in \Omega\mathcal{D}\,(\,\sigma\,)$. Write E for the set of environments.

The semantics of assertions are given by an inductive definition. For each closed type expression σ we define an *interpretation function*

$$[\![\ \ ]\!]_\sigma : \mathcal{M}_\sigma \to [\,E \to \Omega(\mathcal{D}(\,\sigma\,))\,]$$

in the following structural way. For each type σ, define

$$\begin{aligned}
&[\![\,\mathbf{t}\,]\!]_\sigma\rho = \mathcal{D}(\,\sigma\,),\\
&[\![\,\mathbf{f}\,]\!]_\sigma\rho = \emptyset,\\
&[\![\,\varphi \vee \psi\,]\!]_\sigma\rho = [\![\,\varphi\,]\!]_\sigma\rho \cup [\![\,\psi\,]\!]_\sigma\rho,\\
&[\![\,\varphi \wedge \psi\,]\!]_\sigma\rho = [\![\,\varphi\,]\!]_\sigma\rho \cap [\![\,\psi\,]\!]_\sigma\rho,\\
&[\![\,x^\sigma\,]\!]\rho = \rho(\,x^\sigma\,),\\
&[\![\,\mu x^\sigma.\,\varphi(\,x^\sigma\,)\,]\!]_\sigma\rho = \bigcup_{i\in\omega} [\![\,\varphi(\,x^\sigma\,)\,]\!]^i_\sigma\rho,
\end{aligned}$$

where $[\![\,\varphi(\,x^\sigma\,)\,]\!]^0_\sigma\rho = \emptyset$ and, in general,

$$[\![\,\varphi(\,x^\sigma\,)\,]\!]^i_\sigma\rho = [\![\,\varphi(\,x^\sigma\,)\,]\!]_\sigma\rho[x^\sigma \mapsto [\![\,\varphi(\,x^\sigma\,)\,]\!]^{i-1}_\sigma\rho].$$

Here

$$\rho[x^\sigma \mapsto K](y^\tau) = \begin{cases} \rho(y^\tau) & \text{if } x^\sigma \neq y^\tau, \\ K & \text{if } x^\sigma = y^\tau, \end{cases}$$

with $K \in \Omega(\mathcal{D}(\sigma))$.

With respect to type constructions we define

$$\begin{aligned}
[\![\varphi \times \psi]\!]_{\sigma\times\tau}\rho &= \{\langle u, v\rangle \mid u \in [\![\varphi]\!]_\sigma\rho \;\&\; v \in [\![\psi]\!]_\tau\rho\}, \\
[\![\mathit{inl}\,\varphi]\!]_{\sigma+\tau}\rho &= \{\langle 0, u\rangle \mid u \in [\![\varphi]\!]_\sigma\rho \setminus \{\bot_{\mathcal{D}(\sigma)}\}\} \\
&\quad \cup \{x \in \mathcal{D}(\sigma+\tau) \mid \bot_{\mathcal{D}(\sigma)} \in [\![\varphi]\!]_\sigma\rho\}, \\
[\![\mathit{inr}\,\varphi]\!]_{\sigma+\tau}\rho &= \{\langle 1, u\rangle \mid u \in [\![\varphi]\!]_\tau\rho \setminus \{\bot_{\mathcal{D}(\tau)}\}\} \\
&\quad \cup \{x \in \mathcal{D}(\sigma+\tau) \mid \bot_{\mathcal{D}(\tau)} \in [\![\varphi]\!]_\tau\rho\}, \\
[\![\varphi \to \psi]\!]_{\sigma\to\tau}\rho &= \{f \in \mathcal{D}(\sigma) \to \mathcal{D}(\tau) \mid [\![\varphi]\!]_\sigma \subseteq f^{-1}([\![\psi]\!]_\tau\rho)\}, \\
[\![(\varphi)_\bot]\!]_{\sigma_\bot}\rho &= \{\langle 0, u\rangle \mid u \in [\![\varphi]\!]_\sigma\rho\}, \\
[\![\varphi]\!]_{rect.\,\sigma}\rho &= \{\epsilon_\sigma(u) \mid u \in [\![\varphi]\!]_{\sigma[(rect.\,\sigma)\backslash t]}\rho\},
\end{aligned}$$

where, as before, $\epsilon_\sigma : \mathcal{D}(\sigma[(rect.\,\sigma)\backslash t]) \to \mathcal{D}(rect.\,\sigma)$ is the isomorphism arising form the initial solution to the domain equation associated with type $rect.\,\sigma$.

Often the associated type is implicit from the context, so type-subscripts are omitted. We often write $[\![\;\;]\!]_\rho$ for the semantic interpretation. We show that our definition has the desired effect, *i.e.*, $\mu x.\,\varphi(x)$ is the least fixed-point of a semantic function related to $\varphi(x)$.

Definition 5.2 *Let D be a Scott domain and $\Omega(D)$ be the set of Scott open sets of D. A function $F : \Omega(D) \to \Omega(D)$ is continuous if it is monotonic, and for each chain*

$$A_0 \subseteq A_1 \subseteq A_2 \cdots \subseteq A_i \cdots$$

in $\Omega(D)$ we have

$$F(\bigcup_{i\in\omega} A_i) = \bigcup_{i\in\omega} F(A_i).$$

Proposition 5.1 *For each assertion $\varphi(x)$ of type σ and an environment ρ,*

$$\lambda X.\,\Phi_\rho(X) : \Omega(\mathcal{D}(\sigma)) \longrightarrow \Omega(\mathcal{D}(\sigma))$$

is a continuous function, where $\Phi_\rho(A) = [\![\varphi(x)]\!]_{\rho[x\mapsto A]}$, for any $A \in \Omega(\mathcal{D}(\sigma))$.

Proof By structural induction on the assertions. The base cases are trivial.

Every such $\lambda X.\,\Phi_\rho(X)$ is monotonic because free variables are not allowed on the left of $\rightarrow$. For each chain

$$A_0 \subseteq A_1 \subseteq A_2 \cdots \subseteq A_i \cdots$$

in $\Omega(\mathcal{D}(\sigma))$, we have the following equations, which complete the proof.

$$\begin{aligned}
&\lambda X.\,[\![\,\varphi \vee \psi\,]\!]_{\rho[x^\sigma \mapsto X]}(\bigcup_{i\in\omega} A_i)\\
&\quad= \lambda X.\,[\![\,\varphi\,]\!]_{\rho[x^\sigma \mapsto X]}(\bigcup_{i\in\omega} A_i) \cup \lambda X.\,[\![\,\psi\,]\!]_{\rho[x^\sigma \mapsto X]}(\bigcup_{i\in\omega} A_i)\\
&\quad= \bigcup_{i\in\omega} \lambda X.\,[\![\varphi]\!]_{\rho[x^\sigma \mapsto X]}(A_i) \cup \bigcup_{i\in\omega} \lambda X.\,[\![\psi]\!]_{\rho[x^\sigma \mapsto X]}(A_i)\\
&\quad= \bigcup_{i\in\omega} \lambda X.\,[\![\,\varphi \vee \psi\,]\!]_{\rho[x^\sigma \mapsto X]}(A_i),\\
&\lambda X.\,[\![\,\varphi \wedge \psi\,]\!]_{\rho[x^\sigma \mapsto X]}(\bigcup_{i\in\omega} A_i)\\
&\quad= \lambda X.\,[\![\,\varphi\,]\!]_{\rho[x^\sigma \mapsto X]}(\bigcup_{i\in\omega} A_i) \cap \lambda X.\,[\![\,\psi\,]\!]_{\rho[x^\sigma \mapsto X]}(\bigcup_{i\in\omega} A_i)\\
&\quad= (\bigcup_{i\in\omega} \lambda X.\,[\![\varphi]\!]_{\rho[x^\sigma \mapsto X]}(A_i)) \cap (\bigcup_{i\in\omega} \lambda X.\,[\![\psi]\!]_{\rho[x^\sigma \mapsto X]}(A_i))\\
&\quad= \bigcup_{i\in\omega} \lambda X.\,[\![\,\varphi \wedge \psi\,]\!]_{\rho[x^\sigma \mapsto X]}(A_i),\\
&\lambda X.\,[\![\,\mu t.\,\varphi(t,\,x)\,]\!]_{\rho[x^\sigma \mapsto X]}(\bigcup_{i\in\omega} A_i)\\
&\quad= \bigcup_{j\in\omega} \lambda X.\,[\![\,\varphi\,]\!]^j_{\rho[x^\sigma \mapsto X]}(\bigcup_{i\in\omega} A_i)\\
&\quad= \bigcup_{j\in\omega} \bigcup_{i\in\omega} \lambda X.\,[\![\,\varphi\,]\!]^j_{\rho[x^\sigma \mapsto X]}(A_i)\\
&\quad= \bigcup_{i\in\omega} \bigcup_{j\in\omega} \lambda X.\,[\![\,\varphi\,]\!]^j_{\rho[x^\sigma \mapsto X]}(A_i)\\
&\quad= \bigcup_{i\in\omega} \lambda X.\,[\![\,\mu t.\,\varphi(t,\,x)\,]\!]_{\rho[x^\sigma \mapsto X]}(A_i).
\end{aligned}$$

□

As a direct consequence of Proposition 5.1 we have

Proposition 5.2 *For any assertion φ of type σ and any environment ρ,*

$$[\![\,\mu t.\,\varphi\,]\!]_\rho$$

is the least fixed-point of

$$\lambda X.\,[\![\,\varphi\,]\!]_{\rho[x^\sigma \mapsto X]}.$$

5.2 Theorems

The proof system for the mu-calculus consists of proof rules for lifting, sum, product, and function space introduced in Chapter 4. The meta-predicates **P** and **Q** are also used as part of the system.

The following axioms and rules are introduced for μ-assertions.

(μ-axiom) $$Q(\mu x.\, Q(x)) \leq \mu x.\, Q(x)$$

(μ-rule) $$\frac{Q(\varphi) \leq \varphi}{\mu x.\, Q(x) \leq \varphi}$$

These are not all the rules of the μ-calculus yet. We will add to the system two new rules introduced in Section 5.3. For notational convenience we write p or q for closed assertions and $P(x)$, $Q(y)$ for assertions with possible variables indicated in the bracket. Often $\vdash_S A$ means that A can be proved from the system S, or A is a theorem of S. We will often omit $\vdash$ as the context will always provide enough information for that.

Note that the monotonicity rule,

(mono) $$\frac{x \leq y}{Q(x) \leq Q(y)},$$

can be derived from the rest of the rules.

Clearly there is no problem in renaming bound variables. We have, for example,

$$\mu x.\, P(x) = \mu y.\, P(y).$$

This is because $P(\mu y.\, P(y)) = \mu y.\, P(y)$, and one can apply the μ-rule to get

$$\mu x.\, P(x) \leq \mu y.\, P(y).$$

Many useful theorems can now be derived by using the μ-axiom and the μ-rule. The fact that $\mu x.\ Q(x)$ is a fixed-point of Q can be derived from the calculus as stated in Theorem 5.1.

Theorem 5.1

$$Q(\mu x.\, Q(x)) = \mu x.\, Q(x).$$

Proof

$$\begin{aligned}
Q(\mu x.\, Q(x)) &\le \mu x.\, Q(x), && \mu\text{-axiom}\\
Q(Q(\mu x.\, Q(x))) &\le Q(\mu x.\, Q(x)), && \text{monotonicity}\\
\mu x.\, Q(x) &\le Q(\mu x.\, Q(x)), && \mu\text{-rule}\\
\mu x.\, Q(x) &= Q(\mu x.\, Q(x)). && \mu\text{-axiom}
\end{aligned}$$

□

Theorem 5.2

$$\mu x.\, [p \wedge Q(x)] \le \mu x.\, Q(x),$$

where x does not occur free in p.

Proof

$$\begin{aligned}
Q(\mu x.\, Q(x)) &\le \mu x.\, Q(x), && \mu\text{-axiom}\\
p \wedge Q(\mu x.\, Q(x)) &\le \mu x.\, Q(x), && \text{logical rule}\\
\mu x.\, [p \wedge Q(x)] &\le \mu x.\, Q(x). && \mu\text{-rule}
\end{aligned}$$

□

There is the following derived rule which includes Theorem 5.2 as a special case:

$$\frac{P(x) \le Q(x)}{\mu y.\, P(y) \le \mu y.\, Q(y)}.$$

Instantiate x by $\mu y.\, Q(y)$ in the assumption we get

$$P(\mu y.\, Q(y)) \le Q(\mu y.\, Q(y)).$$

But

$$Q(\mu y.\, Q(y)) \le \mu y.\, Q(y)$$

by the μ-axiom; thus

$$\mu y.\, P(y) \le \mu y.\, Q(y)$$

by the μ-rule.

Theorem 5.3 *We have $\mu x.\, Q(x) \le p$ implies*

$$\mu x.\, Q(x) \le \mu x.\, (p \wedge Q(x)).$$

That is,

$$\frac{\mu x.\, Q(x) \le p}{\mu x.\, Q(x) \le \mu x.\, (p \wedge Q(x))}$$

is a derived rule.

Proof

$$\begin{array}{rll}
\mu x.\,(p \wedge Q(x)) & \leq \mu x.\,Q(x), & \text{Theorem 5.2} \\
Q(\mu x.\,p \wedge Q(x)) & \leq \mu x.\,Q(x), & \text{monotonicity and } \mu\text{-axiom} \\
 & \leq p, & \text{assumption} \\
Q(\mu x.\,p \wedge Q(x)) & \leq p \wedge Q(\mu x.\,p \wedge Q(x)), & \text{logical axiom} \\
Q(\mu x.\,p \wedge Q(x)) & \leq \mu x.\,p \wedge Q(x), & \mu\text{-axiom} \\
\mu x.\,Q(x) & \leq \mu x.\,(p \wedge Q(x)). & \mu\text{-rule}
\end{array}$$

□

Theorem 5.4 *We have* $Q(\mu x.\,p \wedge Q(x)) = p$ *implies* $\mu x.\,Q(x) = p$. *In other words,*

$$\frac{Q(\mu x.\,p \wedge Q(x)) = p}{\mu x.\,Q(x) = p}$$

is a derived rule.

Proof

$$\begin{array}{rll}
Q(\mu x.\,p \wedge Q(x)) & = p, & \text{assumption} \\
Q(\mu x.\,p \wedge Q(x)) & = p \wedge Q(\mu x.\,p \wedge Q(x)), & \text{logical rule} \\
 & = \mu x.\,(p \wedge Q(x)), & \text{Theorem 5.1} \\
\mu x.\,Q(x) & \leq \mu x.\,p \wedge Q(x), & \mu\text{-rule} \\
\mu x.\,Q(x) & = \mu x.\,p \wedge Q(x), & \text{Theorem 5.2} \\
Q(\mu x.\,Q(x)) & = Q(\mu x.\,p \wedge Q(x)), & \text{monotonicity} \\
\mu x.\,Q(x) & = p. & \text{assumption}
\end{array}$$

□

Note the same proof goes through by changing $=$ to $\leq$, *i.e.*, we have the derived rule

$$\frac{Q(\mu x.\,p \wedge Q(x)) \leq p}{\mu x.\,Q(x) \leq p}.$$

Some terminologies are necessary for further theorems. Call an assertion $Q(x)$ ***distributive over*** $\vee$ if

$$Q(x \vee y) = Q(x) \vee Q(y)$$

is derivable. Call $Q(x)$ ***distributive over*** $\wedge$ if

$$Q(x \wedge y) = Q(x) \wedge Q(y)$$

is derivable. $Q(x)$ is *distributive* if $Q(x)$ is distributive over both $\vee$ and $\wedge$, and $Q(\mathbf{f}) = \mathbf{f}$. Here x and y are assertion variables.

If $Q(x)$ is distributive over $\vee$ then so is $Q^n(x)$ for any natural number n, where $Q^0(x) = x$, and $Q^i(x) = Q^{i-1}(Q(x))$. This can be checked by mathematical induction. Similarly if $Q(x)$ is distributive over $\wedge$ then so is $Q^n(x)$ for any natural number n, and if $Q(x)$ is distributive then for any natural number n, $Q^n(x)$ is distributive. Note if R distributes over $\vee$ then so does $p \vee R$ and $p \wedge R$ for any closed assertion p; If R distributes over $\wedge$ then so does $p \vee R$ and $p \wedge R$ for any closed assertion p. Evidently, then, if R is distributive then so is $p \wedge R$ for any closed assertion p.

There are assertions which are not distributive over $\vee$. The assertion $q \rightarrow x$ with q non-prime is an example. However, we have

Lemma 5.1 *For any type σ, $Q(x) : \sigma$ distributes over $\wedge$ provided $Q(x)$ does not have any μ-subassertion. If x is the only free variable in $P(x)$, then there exist p, $R(x)$, such that $P(x) = p \vee R(x)$ is derivable from the calculus, where p is closed, and $R(\mathbf{f}) = \mathbf{f}$. Here $P(x)$ is assumed to have no μ-subassertion.*

Proof By inspecting the proof rules one can see that all the assertion constructors *inl*, *inr*, $(\quad)_\perp$, $\rightarrow$, $\times$ distribute over $\wedge$. Moreover, $\vee$, $\wedge$ preserve such distributivity. Hence every such $Q(x)$ distributes over $\wedge$, by an easy structural induction.

To show $\vdash P(x) = p \vee R(x)$, we first use logical axioms to put $P(x)$ into a disjunctive normal form. Let p be the part of the disjunction consisting of all the closed assertions. The rest are disjuncts in which x appear as a free variable. Take this part as $R(x)$, we have $R(\mathbf{f}) = \mathbf{f}$. When dealing with the function space construction, we apply the axiom $\mathbf{f} \rightarrow x = \mathbf{t}$ whenever it is possible, so as to avoid x being considered as free in $\mathbf{f} \rightarrow x$.

□

It is a bore to always indicate which rules are applied. In the remaining of the chapter we shall work more informally.

The following theorem says if the contributor part of a μ-assertion breaks into two parts then the μ-assertion is equivalent to two μ-assertions with the same recursive unit but different contributor parts of the original assertion.

Theorem 5.5 *If $R(x)$ distributes over $\vee$, then*

$$\mu x.\,(p_1 \vee p_2 \vee R(x)) = [\mu x.\, p_1 \vee R(x)] \vee [\mu y.\, p_2 \vee R(y)].$$

Proof Clearly

$$(\mu x.\, p_1 \vee R(x)) \vee (\mu y.\, p_2 \vee R(x)) \leq \mu x.\,(p_1 \vee p_2 \vee R(x)).$$

On the other hand,

$$\begin{aligned} & p_1 \vee p_2 \vee R(\mu x.\, p_1 \vee R(x) \ \vee\ \mu y.\, p_2 \vee R(y)) \\ & = [p_1 \vee R(\mu x.\, p_1 \vee R(x))] \vee [(p_2 \vee R(\mu y.\, p_2 \vee R(y)))] \end{aligned}$$

by assumption. Hence

$$p_1 \vee p_2 \vee R(\mu x.p_1 \vee R(x) \vee \mu y.p_2 \vee R(y)) = \mu x.p_1 \vee R(x) \ \vee\ \mu y.p_2 \vee R(y),$$

which implies, by the μ-rule,

$$\mu z.\, p_1 \vee p_2 \vee R(z) \leq \mu x.\, p_1 \vee R(x) \ \vee\ \mu y.\, p_2 \vee R(y).$$

□

Theorem 5.6 below means under a certain assumption we can expand the contributor of a μ-assertion by applying the recursive unit a number of times and get a new assertion which is equivalent to the original one provided we use a new, appropriate recursive unit.

Theorem 5.6 *Suppose $Q(x)$ is distributive over $\vee$. Then $\forall n > 0$,*

$$\mu x.\ \left[\left(\bigvee_{i=0}^{n} Q^i(p) \right) \vee Q^{n+1}(x) \right] = \mu x.\, p \vee Q(x).$$

Proof We provide a proof for the case $n = 2$. The general case can be shown by mathematical induction. We have

$$\begin{aligned} \mu x.p \vee Q(x) \ & = p \vee Q(p \vee Q(\mu x.\, p \vee Q(x))), \\ & = p \vee Q(p) \vee Q^2(\mu x.\, p \vee Q(x)). \end{aligned}$$

Therefore

$$\mu y.\ p \vee Q(p) \vee Q^2(y) \leq \mu x.\, p \vee Q(x).$$

On the other hand,

$$\begin{aligned} p \vee Q(p) \vee Q^2\,&[p \vee Q(\mu y.\, p \vee Q(p) \vee Q^2(y))] \\ &= p \vee Q(p) \vee Q^2(p) \vee Q^3[\mu y.\, p \vee Q(p) \vee Q^2(y)] \\ &= p \vee Q\,[p \vee Q(p) \vee Q^2(\mu y.\, p \vee Q(p) \vee Q^2(y))] \\ &= p \vee Q\,[\,\mu y.\, p \vee Q(p) \vee Q^2(y)\,]\,. \end{aligned}$$

Hence

$$\mu x.\, p \vee Q(p) \vee Q^2(x) \leq p \vee Q(\mu y.\, p \vee Q(p) \vee Q^2(y)).$$

Now we have

$$\begin{aligned} p \vee Q(\mu x.p \vee Q(p) \vee Q^2(x)) \quad &\leq p \vee Q(p \vee Q(\mu y.p \vee Q(p) \vee Q^2(y))) \\ &\leq p \vee Q(p) \vee Q^2(\mu y.p \vee Q(p) \vee Q^2(y)) \\ &\leq \mu y.p \vee Q(p) \vee Q^2(y). \end{aligned}$$

Therefore

$$\mu x.\, p \vee Q(x) \leq \mu y.\, p \vee Q(p) \vee Q^2(y).$$

□

Theorem 5.7

$$Q(\mu x.\, p \vee Q^n(x)) = \mu x.\, Q(p) \vee Q^n(x)$$

for any $n > 0$, provided that $Q(x)$ is distributive over $\vee$.

Proof We have

$$\begin{aligned} p \vee Q^n(\mu x.\, p \vee Q^n(x)) \quad &= \mu x.\, p \vee Q^n(x), \\ Q(p \vee Q^n(\mu x.\, p \vee Q^n(x))) \quad &= Q(\mu x.\, p \vee Q^n(x)), \\ Q(p) \vee Q^n\,[Q(\mu x.\, p \vee Q^n(x))] \quad &\leq Q(\mu x.\, p \vee Q^n(x)). \end{aligned}$$

Therefore

$$\mu x.\, Q(p) \vee Q^n(x) \leq Q\,[\mu x.\, p \vee Q^n(x)]\,.$$

On the other hand,

$$\begin{aligned} p \vee Q^n\,&[p \vee Q^{n-1}(\mu x.\, Q(p) \vee Q^n(x))] \\ &= p \vee Q^n(p) \vee Q^{2n-1}(\mu x.\, Q(p) \vee Q^n(x)) \\ &= p \vee Q^{n-1}\,[Q(p) \vee Q^n(\mu x.\, Q(p) \vee Q^n(x))] \\ &= p \vee Q^{n-1}(\mu x.\, Q(p) \vee Q^n(x)), \end{aligned}$$

so

$$\mu x.\, p \vee Q^n(x) \leq p \vee Q^{n-1}(\mu x.\, Q(p) \vee Q^n(x)).$$

Therefore

$$\begin{aligned} Q(\mu x.\, p \vee Q^n(x)) &\leq Q(p) \vee Q^n(\mu x.\, Q(p) \vee Q^n(x)), \\ &\leq \mu x.\, Q(p) \vee Q^n(x). \end{aligned}$$

□

The foregoing theorem states that under a certain condition, the effect of applying the recursive unit to the whole μ-assertion is the same as only applying it to the contributor.

By repeated applications of Theorem 5.7 we have $\forall m \geq 0$ and $n > 0$,

$$Q^m(\mu x.\, p \vee Q^n(x)) = \mu x.\, Q^m(p) \vee Q^n(x)$$

provided that $Q(x)$ distributes over $\vee$.

For certain forms of recursive unit with a disjunction of two parts, the μ-assertion is equivalent to a disjunction of two μ-assertions each of which has one part of the original recursive unit. More precisely,

Theorem 5.8 *If $Q(x)$ distributes over $\vee$, then*

$$\begin{aligned} &\mu x.\, p \vee Q^m(x) \vee Q^n(x) \\ &= \textstyle\bigvee_{i=0}^{m-1} [\mu x.\, Q^{i \cdot n}(p) \vee Q^m(x)] \;\vee\; \bigvee_{j=0}^{n-1} [\mu y.\, Q^{j \cdot m}(p) \vee Q^n(y)]. \end{aligned}$$

Proof For any i with $0 \leq i \leq m-1$,

$$\begin{aligned} \mu x.\, Q^{i \cdot n}(p) \vee Q^m(x) &= Q^{i \cdot n}[\mu x.\, p \vee Q^m(x)] \\ &\leq Q^{(i-1) \cdot n}[p \vee Q^n(\mu x.\, p \vee Q^m(x))] \\ &\leq Q^{(i-1) \cdot n}[p \vee Q^n(\mu x.\, p \vee Q^m(x) \vee Q^n(x)) \\ &\qquad \vee Q^m(\mu x.\, p \vee Q^m(x) \vee Q^n(x))] \\ &= Q^{(i-1) \cdot n}[\mu x.\, p \vee Q^m(x) \vee Q^n(x)] \\ &\leq Q^{(i-2) \cdot n} Q^n[\mu x.\, p \vee Q^m(x) \vee Q^n(x)] \\ &\leq Q^{(i-2) \cdot n}[p \vee Q^m(\mu x.\, p \vee Q^m(x) \vee Q^n(x)) \\ &\qquad \vee Q^n(\mu x.\, p \vee Q^m(x) \vee Q^n(x))] \\ &\;\;\vdots \\ &\leq \mu x.\, p \vee Q^m(x) \vee Q^n(x). \end{aligned}$$

Therefore,

$$\bigvee_{i=0}^{m-1} [\mu x.\, Q^{i\cdot n}(p) \vee Q^m(x)] \leq \mu x.\, p \vee Q^m(x) \vee Q^n(x).$$

Similarly,

$$\bigvee_{j=0}^{n-1} [\mu y.\, Q^{j\cdot m}(p) \vee Q^n(y)] \leq \mu x.\, p \vee Q^m(x) \vee Q^n(x).$$

On the other hand, we need to show

$$p \vee Q^m(RHS) \vee Q^n(RHS) \leq RHS,$$

where RHS is an abbreviation for the assertion

$$\left[\mu x.\, \bigvee_{i=0}^{m-1} Q^{i\cdot n}(p) \vee Q^m(x)\right] \vee \left[\mu y.\, \bigvee_{j=0}^{n-1} Q^{j\cdot m}(p) \vee Q^n(y)\right]$$

which is, by Theorem 5.5, equivalent to the right hand side of the original formula.

We show $Q^m(RHS) \leq RHS$; The proof for

$$Q^n(RHS) \leq RHS$$

is similar.

$$Q^m \left[\bigvee_{i=0}^{m-1} [\mu x.\, Q^{i\cdot n}(p) \vee Q^m(x)]\right] \vee \left[\bigvee_{j=0}^{n-1} [\mu y.\, Q^{j\cdot m}(p) \vee Q^n(y)]\right]$$

$$= \bigvee_{i=0}^{m-1} Q^m[\mu x.\, Q^{i\cdot n}(p) \vee Q^m(x)] \vee \bigvee_{j=0}^{n-1} Q^m[\mu y.\, Q_{i\cdot n}(p) \vee Q^m(y)]$$

$$\leq \bigvee_{i=0}^{m-1} [Q^{i\cdot n}(p) \vee Q^m[\mu x.\, Q^{i\cdot n}(p) \vee Q^m(x)]]$$

$$\vee \bigvee_{j=1}^{n-1} [\mu y.\, (Q^{j\cdot m}(p) \vee Q^n(y))] \vee \mu y.\, [Q^{n\cdot m}(p) \vee Q^n(y)]$$

$$\leq \bigvee_{i=0}^{m-1} [\mu x.\, Q^{i\cdot n}(p) \vee Q^m(x)]$$

$$\vee \bigvee_{j=0}^{n-1} [\mu y.\, Q^{j\cdot m}(p) \vee Q^n(y)] \vee \mu y.\, [Q^{n\cdot m}(p) \vee Q^n(y)].$$

However,

$$\begin{aligned} Q^{m\cdot n}(p) \vee Q^n(\mu y.\, p \vee Q^n(y)) &\leq Q^{m\cdot n}(p) \vee p \vee Q^n(\mu y.\, p \vee Q^n(y)) \\ &\leq Q^{m\cdot n}(p) \vee \mu y.\, p \vee Q^n(y) \\ &\leq \mu y.\, p \vee Q^n(y). \end{aligned}$$

Therefore,

$$\mu y.\, [Q^{m\cdot n}(p) \vee Q^n(y)] \leq \mu y.\, p \vee Q^n(y).$$

Hence $Q^m(RHS) \leq RHS$. Now, apply the μ-rule to get

$$\mu x.\, p \vee Q^m(x) \vee Q^n(x) \leq RHS.$$

□

Generalizing Theorem 5.8 we have

$$\begin{aligned} \mu x.\, p \vee Q^{n_1}(x) \;\; & \vee Q^{n_2}(x) \cdots \vee Q^{n_k}(x) \\ = \;\; & \mu x.\, \left[\bigvee_{i=0}^{n_1 - 1} \bigvee_{j=0}^{k} Q^{i\cdot n_j}(p) \right] \vee Q^{n_1}(x) \\ & \vee \mu x.\, \left[\bigvee_{i=0}^{n_2 - 1} \bigvee_{j=0}^{k} Q^{i\cdot n_j}(p) \right] \vee Q^{n_2}(x) \\ & \quad \vdots \\ & \vee \mu x.\, \left[\bigvee_{i=0}^{n_k - 1} \bigvee_{j=0}^{k} Q^{i\cdot n_j}(p) \right] \vee Q^{n_k}(x). \end{aligned}$$

We remark that it is not necessary to work on assertions of a particular form like $\mu x.p \vee P(x)$; We could have worked on $\mu x.\, Q(x)$ directly. However no generality is lost in the way we did because we can always, as a special case, let p be $\mathbf{f}$. This special case is worth noticing for Theorem 5.6, Theorem 5.7, and Theorem 5.8. List below are special cases of these theorems:

$$\mu x.\, \left[(\bigvee_{i=0}^{n} Q^i(\mathbf{f})) \vee Q^{n+1}(x) \right] = \mu x.\, Q(x),$$

$$Q(\mu x.\, Q^n(x)) = \mu x.\, Q(\mathbf{f}) \vee Q^n(x),$$

and

$$\mu x.\, Q^m(x) \vee Q^n(x)$$

$$= \bigvee_{i=0}^{m-1} [\mu x.\, Q^{i\cdot n}(\mathbf{f}) \vee Q^m(x)] \;\vee\; \bigvee_{j=0}^{n-1} [\mu y.\, Q^{j\cdot m}(\mathbf{f}) \vee Q^n(y)],$$

where the same assumption is made about the $Q(x)$'s as in these theorems.

The following theorem is sometimes very useful. It reduces certain forms of nested μ-recursions into a single recursion.

Theorem 5.9

$$\mu x.\,(\mu y.\ P(x,y)) = \mu z.\ P(z,z).$$

Note for $\mu x.\,(\mu y.\ P(x,y))$ to be well-formed the variables x, y must have the same type.

Proof

$$\begin{aligned} P(\mu z.P(z,z),\ \mu z.\ P(z,z)) &= \mu z.\ P(z,z),\\ \mu x.\ P(x,\ \mu z.P(z,z)) &\le \mu z.\ P(z,z),\\ \mu y.\,(\mu x.\ P(x,y)) &\le \mu z.\ P(z,z). \end{aligned}$$

Also,

$$\mu x.\,(\mu y.\ P(x,y)) \le \mu z.\ P(z,z).$$

On the other hand,

$$\begin{aligned} &\mu t.\ P(\mu x.\,[\mu y.\,P(x,y)],\ t)\\ &\qquad = \mu x.\,(\mu y.\,P(x,y)),\\ &P(\mu x.\,[\mu y.\ P(x,y)],\ \mu t.\ P(\mu x.\,[\mu y.\,P(x,y)],t))\\ &\qquad = \mu t.P(\mu x.\,[\mu y.\,P(x,y)],\ t),\\ &P(\mu x.\,[\mu y.\ P(x,y)],\ \mu x.\,[\mu y.\ P(x,y)])\\ &\qquad = \mu x.\,[\mu y.\,P(x,y)]. \end{aligned}$$

Hence

$$\mu z.\ P(z,\ z) \le \mu x.\,[\mu y.\ P(x,\ y)].$$

□

In general we have

$$\mu x_1.\,(\mu x_2.\,(\ \cdots\ \mu x_n.\ P(x_1,x_2,\cdots,x_n))) = \mu x.\ P(x,\ x,\ \cdots,x).$$

As a particular instance of the formula, we have

$$\mu x.\,\mu y.\,(Q(y) \vee P(x)) = \mu z.\,Q(z) \vee P(z).$$

Finally, we have

Theorem 5.10 *For $m > 0$, $n > 0$,*

$$\mu x.(P^m(x) \vee \mu y.p \vee P^n(y)) = \mu x.p \vee P^m(x) \vee P^n(x)$$

provided that $P(x)$ distributes over $\vee$.

Proof It is easy to see that

$$\begin{aligned} & P^m(\mu x.p \vee P^m(x) \vee P^n(x)) \vee \mu y.p \vee P^n(y) \\ & \leq p \vee P^m(\mu x.p \vee P^m(x) \vee P^n(x)) \\ & \qquad \vee P^n(\mu x.p \vee P^m(x) \vee P^n(x)) \vee \mu y.p \vee P^n(y) \\ & \leq \mu x.p \vee P^m(x) \vee P^n(x). \end{aligned}$$

By the μ-rule, we have

$$\mu x.(P^m(x) \vee \mu y.p \vee P^n(y)) \leq \mu x.p \vee P^m(x) \vee P^n(x).$$

On the other hand, by Theorem 5.7,

$$\begin{aligned} & p \vee P^m(\mu x.[P^m(x) \vee \mu y.p \vee P^n(y)]) \\ & \qquad \vee P^n(\mu x.[P^m(x) \vee \mu y.p \vee P^n(y)]) \\ & \leq p \vee \mu x.[P^m(x) \vee P^m(\mu y.p \vee P^n(y))] \\ & \qquad \vee \mu x.[P^m(x) \vee P^n(\mu y.p \vee P^n(y))] \\ & \leq p \vee \mu x.[P^m(x) \vee P^m(\mu y.p \vee P^n(y))] \\ & \qquad \vee \mu x.[P^m(x) \vee \mu y.p \vee P^n(y)] \\ & \leq \mu x.[P^m(x) \vee \mu y.p \vee P^n(y)]. \end{aligned}$$

In the last step we used the fact

$$\mu x.[P^m(x) \vee P^m(\mu y.p \vee P^n(y))] \leq \mu x.[P^m(x) \vee \mu y.p \vee P^n(y)],$$

which follows from the μ-rule by noticing

$$\begin{aligned} & P^m[\mu x.(P^m(x) \vee \mu y.p \vee P^n(y))] \ \vee \ P^m(\mu y.p \vee P^n(y)) \\ & \leq P^m[\mu x.(P^m(x) \vee \mu y.p \vee P^n(y))] \\ & \qquad \vee \mu y.p \vee P^n(y) \ \vee \ P^m(\mu y.p \vee P^n(y)) \\ & \leq \mu x.[P^m(x) \vee \mu y.p \vee P^n(y)] \\ & \qquad \vee P^m(\mu x.[P^m(x) \vee \mu y.p \vee P^n(y)]) \vee \mu y.p \vee P^n(y) \\ & \leq \mu x.[P^m(x) \vee \mu y.p \vee P^n(y)]. \end{aligned}$$

Now applying the μ-rule we get

$$\mu x.\, p \vee P^m(x) \vee P^n(x) \leq \mu x.\, (\, P^m(x) \vee \mu y.\, p \vee P^n(y)\,).$$

□

It follows from Theorem 5.10 that

$$\mu x.\, (\, P^m(x) \vee \mu y.\, p \vee P^n(y)\,) = \mu x.\, (\, P^n(x) \vee \mu y.\, p \vee P^m(y)\,)$$

as long as P distributes over $\vee$, where $m > 0,\ n > 0$.

5.3 Two Special Rules

Let us see what we can derive from the μ-calculus of a particular type, the type of natural numbers $\mathbf{N}_\perp$ specified by $rec\, t.\, \mathbf{1}_\perp + t$. Intuitively $\mathit{inl}\,\mathbf{t}_\perp : \mathbf{N}_\perp$ denotes 'zero', thus we abbreviate it as 0; inr the successor function, and we abbreviate it as s. By Theorem 5.7, we can derive

$$s\ (\mu x.\, [\,0 \vee s^2(x)]\,) = \mu x.\, [\, s\ (\,0\,) \vee s^2(x)\,],$$

which means if we apply the successor to the whole set of even numbers we get the set of odd numbers, as expected.

Many other facts about natural numbers are derivable from the μ-calculus as special instances of the theorems given in the previous section. Listed below are a few of them:

$$\begin{aligned}
&\mu x.\, s(x) = \mathbf{f},\\
&\mu x.\, (0 \vee s^2 x\,) = 0 \vee \mu x.\, s^2(0 \vee s^2 x\,),\\
&\mu x.\, (0 \vee s(0) \vee s^2 x\,) = (\,\mu x.\, 0 \vee s^2 x\,)\ \vee\ (\,\mu x.\, s(0) \vee s^2 x\,),\\
&\mu x.\, (0 \vee s(0) \vee s^2 x\,) = \mu x.\, 0 \vee s\, x.
\end{aligned}$$

Now consider the assertion

$$(\,\mu x.\, 0 \vee s^2\, x\,) \wedge (\,\mu x.\, s(0) \vee s^2\, x\,).$$

It expresses the intersection of even numbers and odd numbers and hence should be the empty set, that is,

$$(\,\mu x.\, 0 \vee s^2\, x\,) \wedge (\,\mu x.\, s(0) \vee s^2\, x\,) = \mathbf{f}.$$

However, this formula does not seem to be derivable from the rules introduced so far (try it!). The reason for this is perhaps that although the μ-axiom and the μ-rule can be used to reason about the least fixed-points expressed in the form $\mu x.P(x)$, they do not necessarily provide the power for reasoning about *equivalences of fixed-points*. A least fixed-point (in the domain) can be expressed in many different forms (in syntax).

As a result of the above analysis, we introduce the following rule, to deal with the interaction of conjunction and the μ-construction.

$$(\wedge - \mu) \qquad \frac{p \wedge (\mu x.\, q \vee R(x)) \leq R(p) \qquad R(x) \text{ distributive}}{p \wedge (\mu x.\, q \vee R(x)) \leq \mu x.\, p \wedge (q \vee R(x))}.$$

This rule supplies a way to manipulate assertions like $\mu x.\, p \wedge (q \vee R(x))$ with p closed, which no other theorem so far deals with. As

$$p \wedge (\mu x.\, q \vee R(x)) = \mu x.\, p \wedge (q \vee R(x))$$

is clearly not valid, some conditions are needed for it to hold. The $(\wedge - \mu)$ rule says that R is distributive and

$$p \wedge (\mu x.\, q \vee R(x)) \leq R(p)$$

are sufficient for it to hold.

We check the soundness of this rule by mathematical induction, to show that for all n, for all environment ρ,

$$\bigcup_{0 \leq i \leq n} [\![p \wedge R^i(q)]\!]_\rho \subseteq [\![\mu x.\, p \wedge (q \vee R(x))]\!]_\rho$$

provided that

$$\forall \rho.\ [\![p \wedge \mu x.\, q \vee R(x)]\!]_\rho \subseteq [\![R(p)]\!]_\rho.$$

Here, $R^0(x)$ is an abbreviation for x. Our proof relies on the soundness of the rest of the μ-calculus, of course. The base case $n = 1$ is trivial. For the induction step, first note that

$$\bigcup_{1 \leq i \leq n+1} [\![p \wedge R^i(q)]\!]_\rho \subseteq [\![p \wedge \mu x.\, q \vee R(x)]\!]_\rho.$$

By assumption, therefore,

$$\bigcup_{1 \leq i \leq n+1} [\![p \wedge R^i(q)]\!]_\rho = [\![R(p)]\!]_\rho \cap \bigcup_{1 \leq i \leq n+1} [\![p \wedge R^i(q)]\!]_\rho.$$

Now

$$
\begin{aligned}
&\bigcup_{0\le i\le n+1}[\![p\wedge R^i(q)]\!]_\rho \\
&= [\![p\wedge q]\!]_\rho \cup ([\![p]\!]_\rho \cap \bigcup_{1\le i\le n+1}[\![R^i(q)]\!]_\rho\,)\,) \\
&= [\![p\wedge q]\!]_\rho \cup ([\![p]\!]_\rho \cap [\![R(p)]\!]_\rho \cap \bigcup_{1\le i\le n+1}[\![R^i(q)]\!]_\rho\,)\,) \\
&= [\![p\wedge q]\!]_\rho \cup ([\![p]\!]_\rho \cap [\![R(\bigvee_{0\le i\le n} p\wedge R^i(q)\,)]\!]_\rho\,) \\
&\subseteq [\![p\wedge q]\!]_\rho \cup (\, [\![p]\!]_\rho \cap [\![R(\,\mu x.\, p\wedge (q\vee R(x))\,)]\!]_\rho\,) \\
&= [\![\mu x.\, p\wedge (q\vee R(x))]\!]_\rho .
\end{aligned}
$$

Note in the second last step we used the induction hypothesis

$$\bigcup_{0\le i\le n} [\![p\wedge R^i(q)]\!]_\rho \subseteq [\![\mu x.\, p\wedge (q\vee R(x))]\!]_\rho,$$

and the monotonicity, distributivity of $R(x)$.

Another rule, which is clearly sound, is necessary for sum:

$$(\neg + \mu) \qquad \frac{\mathbf{Q}(P(\mathbf{f})) \qquad \mathbf{Q}(Q(\mathbf{f}))}{\mathit{inl}\,(\mu x.\, P(x)) \wedge \mathit{inr}\,(\mu y.\, Q(y)) \le \mathbf{f}}$$

Note $\mathbf{Q}$ is a meta-predicate to capture a class of assertions not equivalent to $\mathbf{t}$. Thus for $\mathbf{Q}(P(\mathbf{f}))$ to hold, $P(\mathbf{f})$ must be in $\mathcal{A}_\sigma$ for some closed type σ. Therefore when we write $\mathbf{Q}(P(\mathbf{f}))$ and $\mathbf{Q}(Q(\mathbf{f}))$, we mean, in particular, that $P(x)$ and $Q(x)$ do not have μ-subassertions themselves, and x is the only possible free variable in P and Q.

These two rules complete the proof system for the μ-calculus.

The intersection of the set of even numbers and the set of odd numbers can now be proven to be empty by using the two rules just introduced, as well as other rules of the μ-calculus. Clearly $0\wedge s(0)=\mathbf{f}$. By Theorem 5.7,

$$s^2(\,\mu x.\, s(0)\vee s^2\, x\,) = s(\,\mu x.\, s^2 0\vee s^2\, x\,).$$

We have $\mathbf{Q}(\mathbf{t}_\perp)$ and $\mathbf{Q}(s^2 0)$. By $(\neg + \mu)$,

$$0\wedge s(\,\mu x.\, s^2 0\vee s^2\, x\,) = \mathbf{f}.$$

Therefore

$$0\wedge s^2(\,\mu x.\, s(0)\vee s^2\, x\,) = \mathbf{f}.$$

Similarly,

$$s(0)\wedge s^2(\,\mu x.\, 0\vee s^2\, x\,) = \mathbf{f}.$$

These imply

$$\begin{aligned}&(\,\mu x.\,0 \vee s^2\,x\,) \wedge (\,\mu x.\,s(0) \vee s^2\,x\,)\\&= (\,0 \vee s^2(\,\mu x.\,0 \vee 0^2\,x\,)\,) \wedge (\,s(0) \vee s^2(\,\mu x.\,s(0) \vee s^2\,x\,)\,)\\&= s^2[(\,\mu x.\,0 \vee s^2\,x\,) \wedge (\,\mu x.\,s(0) \vee s^2\,x\,)]\\&\leq s^2(\,\mu x.\,0 \vee s^2\,x\,).\end{aligned}$$

Now, by taking p to be $\mu x.\,0 \vee s^2\,x$, q to be $s(0)$, and $R(x)$ to be $s^2(x)$ in $(\wedge - \mu)$, we get

$$\begin{aligned}&(\,\mu x.\,0 \vee s^2\,x\,) \wedge (\,\mu x.\,s(0) \vee s^2\,x\,)\\&\leq \mu y.[\,(\,\mu x.\,0 \vee s^2\,x\,) \wedge (\,s(0) \vee s^2\,y\,)\,]\\&\leq \mu y.[\,(\,\mu x.\,0 \vee s^2\,x\,) \wedge s^2\,y\,]\\&\leq \mathbf{f}.\end{aligned}$$

The kind of formulae that can be derived by using $(\wedge - \mu)$ (and other rules) is described by

Theorem 5.11 *Suppose*

$$p \wedge (\mu x.\,q \vee R(x)) = \mathbf{f}, \quad q \wedge (\mu x.\,p \vee R(x)) = \mathbf{f}$$

are theorems, where $R(x)$ is distributive. Then

$$(\mu x.\,p \vee R(x)) \wedge (\mu x.\,q \vee R(x)) = \mathbf{f}$$

is also a theorem.

Proof We have

$$\begin{aligned}&(\mu x.\,p \vee R(x)) \wedge (\mu x.\,q \vee R(x))\\&= (\,p \vee R(\mu x.\,p \vee R(x))\,) \wedge (\,q \vee R(\mu x.\,q \vee R(x))\,)\\&= R[\,(\mu x.\,p \vee R(x)) \wedge (\mu x.\,q \vee R(x))\,]\\&\leq R(\mu x.\,p \vee R(x)).\end{aligned}$$

By $(\wedge - \mu)$,

$$\begin{aligned}&(\mu x.\,p \vee R(x)) \wedge (\mu x.\,q \vee R(x))\\&\leq \mu y.\,(\mu x.\,p \vee R(x)) \wedge (\,q \vee R(y)\,)\\&\leq \mu y.\,(\mu x.\,p \vee R(x)) \wedge R(y)\\&\leq \mu y.\,R(y)\\&= \mathbf{f}.\end{aligned}$$

□

We remark that the purpose of the rule $(\wedge - \mu)$ is to derive similar results like Theorem 5.11. It is this theorem that is directly used in the proof of completeness of the μ-calculus of integers. Hence we could have done differently, introducing the result of Theorem 5.11 as a new rule.

We can now reduce certain conjunctions of μ-assertions to a single μ-assertion as in

$$(\mu x.\, 0 \vee s^2 x) \wedge (\mu x.\, 0 \vee s^3 x\ ) = \mu x.\, 0 \vee s^6 x.$$

The general pattern of such formulae is captured by Theorem 5.12.

Theorem 5.12 *Suppose $Q(x)$ is distributive and*

$$p \wedge Q^i(\ \mu x.\, p \vee Q^{(m,n)}(x)\ ) = \mathbf{f}$$

for $0 < i < (m,n)$ with $m > 0$, $n > 0$, where (m,n) is the least common multiple of m and n. Then

$$(\mu x.\, p \vee Q^m(x)) \wedge (\mu y.\, p \vee Q^n(y)) = \mu z.\, p \vee Q^{(m,n)}(z).$$

Proof Suppose $m \neq n$. By Theorem 5.6 we have

$$\mu z.\, p \vee Q^{(m,n)}(z) \leq \mu x.\, p \vee Q^m(x),$$

$$\mu z.\, p \vee Q^{(m,n)}(z) \leq \mu y.\, p \vee Q^n(y).$$

Therefore,

$$\mu z.\, p \vee Q^{(m,n)}(z) \leq (\mu x.\, p \vee Q^m(x)) \wedge (\mu y.\, p \vee Q^n(y)).$$

On the other hand, let $(m,\, n) = sm = tn$, for some $s > 0$, $t > 0$. By Theorem 5.6 again,

$$\mu x.\, p \vee Q^m(x) = \mu x.\ \textstyle\bigvee_{i=0}^{s-1} Q^{i \cdot m}(p) \vee Q^{(m,\, n)}(x),$$

$$\mu x.\, p \vee Q^n(y) = \mu y.\ \textstyle\bigvee_{j=0}^{t-1} Q^{j \cdot n}(p) \vee Q^{(m,\, n)}(y).$$

Hence

$$\begin{aligned}
&[\, \mu x.\, p \vee Q^m(x)\,] \wedge [\, \mu y.\, p \vee Q^n(y)\,] \\
&= [\, \textstyle\bigvee_{i=0}^{s-1}(\ \mu x.\ Q^{i \cdot m}(p) \vee Q^{(m,\, n)}(x)\,)\,] \\
&\qquad \wedge [\ \textstyle\bigvee_{j=0}^{t-1}(\ \mu y.\ Q^{j \cdot n}(p) \vee Q^{(m,\, n)}(y)\ )\,].
\end{aligned}$$

Clearly it is enough to show that, for $0 < i < s$, $0 < j < t$,

$$[\,\mu x.\ Q^{i\cdot m}(p) \vee Q^{L}(x)\,] \wedge [\,\mu y.\ Q^{j\cdot n}(p) \vee Q^{L}(y)\,] = \mathbf{f}.$$

For that purpose note we have

$$\begin{aligned}
&[\,\mu x.\ Q^{i\cdot m}(p) \vee Q^{(m,n)}(x)\,] \wedge [\,\mu y.\ Q^{j\cdot n}(p) \vee Q^{(m,n)}(y)\,] \\
&= [\,Q^{i\cdot m}(p) \vee Q^{(m,n)}(\,\mu x.\ Q^{i\cdot m}(p) \vee Q^{(m,n)}(x)\,)\,] \\
&\quad \wedge [\,Q^{j\cdot n}(p) \vee Q^{(m,n)}(\,\mu y.\ Q^{j\cdot n}(p) \vee Q^{(m,n)}(y)\,)\,] \\
&= Q^{(m,n)}\,[\,(\,\mu x.\ Q^{i\cdot m}(p) \vee Q^{(m,n)}(x)\,) \wedge (\,\mu y.\ Q^{j\cdot n}(p) \vee Q^{(m,n)}(y)\,)\,] \\
&\quad \vee [\,Q^{i\cdot m}(p) \wedge Q^{(m,n)}(\,\mu y.\ Q^{j\cdot n}(p) \vee Q^{(m,n)}(y)\,)\,] \\
&\quad \vee [\,Q^{j\cdot n}(p) \wedge Q^{(m,n)}(\,\mu x.\ Q^{i\cdot m}(p) \vee Q^{(m,n)}(x)\,)\,].
\end{aligned}$$

The term $Q^{i\cdot m}(p) \wedge Q^{j\cdot n}(p)$ is equivalent to $\mathbf{f}$, from the fact that $im \neq jn$ and the assumption. However, implied by the assumption we also have

$$Q^{i\cdot m}(p) \wedge Q^{(m,n)}(\,\mu y.\ Q^{j\cdot n}(p) \vee Q^{(m,n)}(y)\,) = \mathbf{f},$$

$$Q^{j\cdot n}(p) \wedge Q^{(m,n)}(\,\mu x.\ Q^{i\cdot m}(p) \vee Q^{(m,n)}(x)\,) = \mathbf{f}.$$

Hence

$$\begin{aligned}
&[\,\mu x.\ Q^{i\cdot m}(p) \vee Q^{(m,n)}(x)\,] \wedge [\,\mu y.\ Q^{j\cdot n}(p) \vee Q^{(m,n)}(y)\,] \\
&= Q^{(m,n)}(\,[\,\mu x.\ Q^{i\cdot m}(p) \vee Q^{(m,n)}(x)\,] \wedge [\,\mu y.\ Q^{j\cdot n}(p) \vee Q^{(m,n)}(y)\,]\,) \\
&\leq Q^{(m,n)}(\,[\,\mu x.\ Q^{i\cdot m}(p) \vee Q^{(m,n)}(x)\,]\,).
\end{aligned}$$

Applying $(\wedge - \mu)$, we get

$$\begin{aligned}
&[\,\mu x.\ Q^{i\cdot m}(p) \vee Q^{(m,n)}(x)\,] \wedge [\,\mu y.\ Q^{j\cdot n}(p) \vee Q^{(m,n)}(y)\,] \\
&\leq \mu y.\ [\,(\,\mu x.\ Q^{i\cdot m}(p) \vee Q^{(m,n)}(x)\,) \wedge (\,Q^{j\cdot n}(p) \vee Q^{L}(y)\,)\,] \\
&\leq \mu y.\ [\,(\,\mu x.\ Q^{i\cdot m}(p) \vee Q^{(m,n)}(x)\,) \wedge Q^{(m,n)}(y)\,] \\
&\leq \mathbf{f}.
\end{aligned}$$

Note we have used the fact

$$(\,\mu x.\ Q^{i\cdot m}(p) \vee Q^{(m,n)}(x)\,) \wedge Q^{j\cdot n}(p) = \mathbf{f},$$

deduced from

$$\mu x.\ Q^{i\cdot m}(p) \vee Q^{(m,n)}(x) = Q^{i\cdot m}(p) \vee Q^{L}(\,\mu x.\ Q^{i\cdot m}(p) \vee Q^{(m,n)}(x)\,),$$

$$Q^{i\cdot m}(p) \wedge Q^{j\cdot n}(p) = \mathbf{f},$$

and

$$Q^{j\cdot n}(p) \wedge Q^{(m,n)}(\ \mu x.\ Q^{i\cdot m}(p) \vee Q^{(m,n)}(x)\ ) = \mathbf{f},$$

all follow from the assumption.

□

5.4 Soundness and Completeness

The soundness of the μ-calculus is easy to establish. We have

Theorem 5.13 *Suppose φ, ψ are assertions of type σ. If $\varphi \leq \psi$, then for any environment ρ, $[\![\varphi]\!]\rho \subseteq [\![\psi]\!]\rho$.*

The proof goes by inspecting that all the axioms are valid and rules sound. In the rest of the section we show that the μ-calculus of the integers $\mathbf{N}_\perp$ is complete. To achieve completeness we put two syntactic restrictions on the assertions:

1. To form the assertion $\mu x.\, P(x)$,
   x is required to be the only possible free variable in P, and
2. To form a conjunction $\varphi_0 \wedge \varphi_1$,
   φ_0 and φ_1 are required to be closed.

As we will see in the next section, fortunately, these restrictions do not affect the expressive power of the μ-calculus for integers.

Definition 5.3 *A closed assertion φ of $\mathbf{N}_\perp$ is called a normal form if for some finite index I,*

$$\varphi \equiv \bigvee_{k \in I} \mu x.\, p_k \vee s^{j_k}(x),$$

where p_k's are either of the form $s^{i_k}\,0$, or $\mathbf{t}$, or $\mathbf{f}$, and i_k, j_k are natural numbers.

When an assertion is provably equivalent to a normal form we say that this assertion *has a normal form*. The following is a key theorem for completeness.

Theorem 5.14 *(The First Normal Form Theorem) Every closed assertion of $\mathbf{N}_\perp$ has a normal form.*

Proof By structural induction. We show that the property of ***having a normal form*** is preserved by all possible ways of forming new assertions (under the two syntactic restrictions). It is obvious that if φ_0 and φ_1 are in normal forms then so is $\varphi_0 \vee \varphi_1$. Thus disjunction preserves normal forms. It is easy to see that, by Theorem 5.7, the operator s preserves normal forms.

To deal with conjunction, let φ_0 and φ_1 be in their normal forms. $\varphi_0 \wedge \varphi_1$ has a normal form if every

$$(\ \mu x.\ s^i\, 0 \vee s^j(x)\ ) \wedge (\ \mu x.\ s^m\, 0 \vee s^n(x)\ )$$

has a normal form for all natural numbers i, j, m, n. By applying Theorem 5.6 first and then Theorem 5.5 we can raise the power of the recursive unit so that the μ-assertions above both have the same power for the recursive units. Therefore we can assume $j = n$. Suppose $i \neq m$, and, without loss of generality, $i < m$. We have

$$\begin{aligned}&(\ \mu x.\ s^i\, 0 \vee s^n(x)\ ) \wedge (\ \mu x.\ s^m\, 0 \vee s^n(x)\ )\\ &= s^i\ [\ (\ \mu x.\ 0 \vee s^n(x)\ ) \wedge (\ \mu x.\ s^{m-i}\, 0 \vee s^n(x)\ )\ ].\end{aligned}$$

Thus it is sufficient to show that

$$(\ \mu x.\ 0 \vee s^n(x)\ ) \wedge (\ \mu x.\ s^k\, 0 \vee s^n(x)\ )$$

has a normal form, where $k > 0$.

If $(\ \mu x.\ 0 \vee s^n(x)\ ) \wedge\ s^k\, 0 = \mathbf{f}$, by Theorem 5.11

$$(\ \mu x.\ 0 \vee s^n(x)\ ) \wedge (\ \mu x.\ s^k\, 0 \vee s^n(x)\ ) = \mathbf{f}$$

since it follows from $(\neg + \mu)$ that, with $k > 0$,

$$0 \wedge (\ \mu x.\ s^k\, 0 \vee s^n(x)\ ) = \mathbf{f}.$$

Otherwise

$$(\ \mu x.\ 0 \vee s^n(x)\ ) \wedge\ s^k\, 0 = s^k\, 0$$

and k must be a multiple of n. Let t be the least number such that $t \cdot n = k$. By virtue of Theorem 5.6 and Theorem 5.5 again, we have

$$\mu x.\ 0 \vee s^n(x) = \bigvee_{0 \le i \le t} \mu x.\ s^{ni} 0 \vee s^{(t+1)\cdot n}(x).$$

By Theorem 5.11,

$$(\ \mu x.\ s^{ni}0 \vee s^{(t+1)\cdot n}(x)\ ) \wedge (\ \mu x.\ s^k\, 0 \vee s^n(x)\ ) = \mathbf{f}$$

when $i < t$. Therefore

$$\begin{aligned}&(\ \mu x.\ 0 \vee s^n(x)\ ) \wedge (\ \mu x.\ s^k\, 0 \vee s^n(x)\ )\\ &= (\ \mu x.\ s^k 0 \vee s^{(t+1)\cdot n}(x)\ ) \wedge (\ \mu x.\ s^k\, 0 \vee s^n(x)\ ).\end{aligned}$$

We can then use Theorem 5.12 to get a normal form.

To show $\mu x.\ P(x) : \mathbf{N}_\perp$ has a normal form we first transform $P(x)$ into a disjunctive normal form

$$P(x) = p \vee \bigvee_{0 \le i \le n} s^{k_i}(x),$$

where p is closed. This is possible because of the two syntactic restrictions imposed. Applying the generalized versions of Theorem 5.8 and Theorem 5.5 we get

$$\mu x.\ P(x) = \bigvee_{0 \le i \le n} \mu x.\, p_i \vee s^{k_i}(x),$$

where p_i's can be assumed to be already of the form

$$\mu x.\ s^k 0 \vee s^m(x).$$

It remains to check that assertions of the form

$$\mu x.\ (\mu y.\ s^k 0 \vee s^j(y)\ ) \vee s^m(x)$$

have normal forms. But by Theorem 5.10,

$$\begin{aligned}&\mu x.\ (\mu y.\ s^k 0 \vee s^j(y)\ ) \vee s^m(x)\\ &= \mu x.\ s^k 0 \vee s^j(x) \vee s^m(x).\end{aligned}$$

It follows from Theorem 5.8 that

$$\mu x.\ s^k 0 \vee s^j(x) \vee s^m(x)$$

has a normal form. Hence $\mu x.\ P(x)$ has a normal form.

□

Note that by Theorem 5.6 it is possible to transform any normal form into an assertion

$$p \vee (\mu x.\ p_0 \vee s^n(x)),$$

where p and p_0 are assertions of $\mathcal{A}_{\mathbf{N}_\perp}$ in their prime normal forms, which do not have μ-subassertions, neither free variables. We can further assume that for each disjuncts $s^k 0$ of p_0, $k < n$. This actually leads to another normal form.

Theorem 5.15 *(The Second Normal Form Theorem) Let $\varphi : \mathbf{N}_\perp$. Then it is derivable that*

$$\varphi = p \vee (\mu x.\, p_0 \vee s^n(x)),$$

where p and p_0 are prime normal forms of $\mathcal{A}_{\mathbf{N}_\perp}$ and for each disjuncts $s^k 0$ of p_0, $k < n$.

Although the first normal form is more intuitive, it is the second normal form that is used in the proof of completeness. Following the notation introduced before, we write $\models \varphi \leq \psi$ when $\forall \rho.\, [\![\varphi]\!]_\rho \subseteq [\![\varphi]\!]_\rho$.

Theorem 5.16 *Suppose φ and ψ are assertions of $\mathbf{N}_\perp$, in their second normal forms. Then $\models \varphi \leq \psi$ implies $\vdash \varphi \leq \psi$.*

Proof Suppose

$$\varphi \equiv p \vee (\mu x.\, p_0 \vee s^n(x)), \qquad \psi \equiv q \vee (\mu x.\, q_0 \vee s^m(x)),$$

and $\models \varphi \leq \psi$. Since $\models \varphi \leq \psi$, it must be possible to unwind ψ a number of times so that $\models p \leq q'$, where q' is the closed part of the resulting second normal form. Because of the completeness of the propositional domain logic (Theorem 4.2), we have $\vdash p \leq q'$. This implies $\vdash p \leq \psi$. Now by virtue of Theorem 5.6 we may assume, without loss of generality, $n = m$. Thus

$$\models \mu x.\, p_0 \vee s^n(x) \leq q \vee (\mu x.\, q_0 \vee s^n(x)).$$

It is sufficient to prove that $\models p_0 \leq q_0$. Suppose it were not the case. Then for some disjuncts $s^k 0$ of p_0, $[\![s^k 0]\!] \neq [\![s^i 0]\!]$ for any disjuncts $s^i 0$ of q_0. This implies $[\![s^{k+ni} 0]\!] \not\subseteq [\![s^{ni}(q_0)]\!]$ for any $i \geq 0$. But q can only represent a compact open, hence finite set of natural numbers. Therefore

$$[\![\mu x.\, p_0 \vee s^n(x)]\!] \not\subseteq [\![q \vee (\mu x.\, q_0 \vee s^n(x))]\!],$$

a contradiction.

□

Immediately there is the

Corollary 5.1 *The μ-calculus for integers consisting of the μ-axiom, the μ-rule, $(\wedge - \mu)$, $(\neg + \mu)$, and those rules for $\mathcal{A}_{\mathbf{N}_\perp}$ is complete.*

5.5 The Definable Subsets of $\mathbf{N}_\perp$

The expressiveness of the μ-calculus follows straightforward from the normal form theorems. From the first normal form we know that the open sets expressible by the assertions of $\mathbf{N}_\perp$ are $\emptyset$, $\mathbf{N}_\perp$ and all sets of the form

$$N_0 \cup \bigcup_{0<i<n} \{ m_i + k \cdot n_i \mid k \in \omega \},$$

where m_i, n_i are natural numbers and N_0 is a finite set of natural numbers. The second normal suggests that the expressible sets can also be put in the form

$$N_0 \cup \bigcup_{k \in \omega} (k \cdot n_0 + N_1),$$

where n_0 is a natural number, N_0 and N_1 are finite sets of natural numbers, and

$$k + N =^{\text{def}} \{ k + n \mid n \in N \}$$

for a natural number k and a set N.

Of course the above expressiveness results are for the assertions under the two syntactic restrictions. Do these restrictions really impose any limitation on the expressive power? The answer is no. This is because we have the following result on the definable subsets of $\mathbf{N}_\perp$. It shows that without any syntactic restriction, the assertions represent exactly the same class of open sets. This result agrees with the one mentioned in [Sh87] for the expressive power of an open set language. However the proof offered in [Sh87] for this particular result does not seem to be adequate. We present a proof of the following theorem in the rest of the section.

Theorem 5.17 *The sets of natural numbers expressible by closed assertions of $\mathbf{N}_\perp$ are of the form*

$$N_0 \cup \bigcup_{k \in \omega} (k \cdot n_0 + N_1),$$

where n_0 is a natural number, N_0 and N_1 are finite sets of natural numbers. The two syntactic restrictions do not affect the expressive power of the μ-calculus for integers.

We remark that it makes no difference if we require the numbers in N_0 and N_1 be no greater than n_0 in the above set of integers. For convenience we call this kind of sets *periodical.* We introduce two interesting lemmas first, which are crucial to the proof of Theorem 5.17.

Definition 5.4 *An assertion* $\varphi(x) : \mathbf{N}_\perp$ *is called a linear shift if*

$$[\![\varphi(p+q)]\!] = p + [\![\varphi(q)]\!]$$

for all closed assertions p, q *of* $\mathbf{N}_\perp$ *such that* $\perp \notin [\![p]\!]$ *and* $\perp \notin [\![q]\!]$.

When writing

$$[\![\varphi(p+q)]\!] = p + [\![\varphi(q)]\!]$$

we have used p, q both as syntactic and as semantic objects. In a more rigorous form, we should rewrite it as

$$[\![\varphi(x)]\!]_{\rho[x \mapsto [\![p]\!]+[\![q]\!]]} = [\![p]\!] + [\![\varphi(q)]\!]_\rho,$$

with $[\![p]\!] + [\![q]\!]$ defined to be the set of natural numbers

$$\{k \mid k = i + j \text{ for some } i \in [\![p]\!],\ j \in [\![q]\!]\ \}.$$

Since this is too clumsy, we prefer to use the informal notation.

As a special case, a linear shift must satisfy

$$[\![\varphi(p)]\!] = p + [\![\varphi(0)]\!],$$

since $[\![p]\!] + 0 = [\![p]\!]$. A closed assertion is not a linear shift unless it is equivalent to $\mathbf{f}$.

Lemma 5.2 *Every assertion* $\varphi(x) : \mathbf{N}_\perp$ *with* $\varphi(\mathbf{f}) = \mathbf{f}$ *is a linear shift.*

Proof We use induction on the structure of φ. Clearly every such assertion, if closed, is a linear shift since $A + \emptyset = \emptyset$. It is clear that x is a linear shift. It is routine to check that $\wedge$, $\vee$ transform linear shifts into a linear shift, noticing that

$$(R + A) \cup (R + B) = R + (A \cup B)$$

and

$$(R + A) \cap (R + B) = R + (A \cap B)$$

for all sets A, B, and R of natural numbers. The successor constructor s also preserves linear shifts. Suppose $[\![\varphi(p+q)]\!] = p + [\![\varphi(q)]\!]$. Then

$$\begin{aligned}[\![s(\varphi(p+q))]\!] &= s(p + [\![\varphi(q)]\!]) \\ &= p + [\![\varphi(q)]\!] + 1 \\ &= p + [\![s(\varphi(q))]\!].\end{aligned}$$

It remains to check that $\mu y.\,\varphi(x, y)$ is a linear shift provided that $\varphi(x, y)$ is. However, note that

$$\begin{aligned}[\![\mu y.\,\varphi(p+q,\, y)]\!] &= \bigcup_{i\in\omega}[\![\varphi^i(p+q,\, \mathbf{f})]\!] \\ &= \bigcup_{i\in\omega}(p + [\![\varphi^i(q,\, \mathbf{f})]\!]) \\ &= p + \bigcup_{i\in\omega}[\![\varphi^i(q,\, \mathbf{f})]\!] \\ &= p + [\![\mu y.\,\varphi(q,\, y)]\!].\end{aligned}$$

Hence $\mu y.\,\varphi(x, y)$ is indeed a linear shift. Here we have used the fact that linear shifts are closed under composition.

□

Although all strict assertions of $\mathbf{N}_\perp$ are linear shifts, as stated in Lemma 5.2, it does not imply anything about the distributivity of $\varphi(x)$. In fact one can find strict assertions that are neither distributive over $\wedge$, nor distributive over $\vee$.

Lemma 5.3 *Suppose $\mu x.\ \varphi(x) : \mathbf{N}_\perp$ is an assertion such that $[\![\varphi(\mathbf{f})]\!]$ is periodical. Then there is a strict $\psi(x) : \mathbf{N}_\perp$ (hence a linear shift) such that*

$$[\![\mu x.\,\varphi(x)]\!] = [\![\mu x.\,\varphi(\mathbf{f}) \vee \psi(x)]\!].$$

Proof First note that the difference of two periodical sets

$$M_0 \cup \bigcup_{k\in\omega}(k\cdot n + M_1) \quad \text{and} \quad N_0 \cup \bigcup_{k\in\omega}(k\cdot n + N_1)$$

is the set

$$(M_0 \setminus N_0) \cup \bigcup_{k\in\omega}(k\cdot n + (M_1 \setminus N_1)).$$

By Theorem 5.6 we know that the difference of two periodical sets is periodical in general. In particular, the set $\omega \setminus [\![\varphi(\mathbf{f})]\!]$ is periodical. Hence for some $\beta : \mathbf{N}_\perp$,

$$\omega \setminus [\![\varphi(\mathbf{f})]\!] = [\![\beta]\!].$$

We clearly have, by the monotonicity of $\varphi(x)$,

$$\begin{aligned}
[\![\varphi(x)]\!] &= [\![\varphi(\mathbf{f})]\!] \cup ([\![\varphi(x)]\!] \setminus [\![\varphi(\mathbf{f})]\!]) \\
&= [\![\varphi(\mathbf{f})]\!] \cup ([\![\varphi(x)]\!] \cap (\omega \setminus [\![\varphi(\mathbf{f})]\!])) \\
&= [\![\varphi(\mathbf{f})]\!] \cup ([\![\varphi(x)]\!] \cap [\![\beta]\!]).
\end{aligned}$$

This analysis shows that we can let $\psi(x)$ be $\beta \wedge \varphi(x)$.

□

Proof of Theorem 5.17 By structural induction. It is easy to see that s, $\vee$, and $\wedge$ preserve periodical sets, since these cases are well taken care of by Theorem 5.5, Theorem 5.6, Theorem 5.7, and Theorem 5.12. We now show that every μ-assertion represents a periodical set provided its sub-expressions do. By Lemma 5.3, it is enough to consider μ-assertions of the form

$$\mu x.\, p \vee \varphi(x)$$

with $\varphi(\mathbf{f}) = \mathbf{f}$. Such a $\varphi(x)$, by Lemma 5.2, is a linear shift. The property implied by a linear shift makes it possible to show, by mathematical induction that

$$\begin{aligned}
[\![\mu x.\, p \vee \varphi(x)]\!] &= [\![p]\!] \\
&\quad \cup([\![p]\!] + [\![\varphi(0)]\!]) \\
&\quad \cup([\![p]\!] + [\![\varphi(0)]\!] + [\![\varphi(0)]\!]) \\
&\quad \cup([\![p]\!] + [\![\varphi(0)]\!] + [\![\varphi(0)]\!] + [\![\varphi(0)]\!]) \\
&\quad \vdots
\end{aligned}$$

The hypothesis of the structural induction implies that both $[\![p]\!]$ and $[\![\varphi(0)]\!]$ are periodical. One can then deduce that $[\![\mu x.\, p \vee \varphi(x)]\!]$ is also periodical, using the above formula.

□

Part II

Stable Domains

Chapter 6

Categories

Part II is devoted to the development of logical frameworks for stable domains (dI-domains). There are many categories of domains with stable domains as objects and certain kinds of stable functions as morphisms. But they are scattered in the literature. To provide a background knowledge for the work of Part II, we give a survey of such categories (some of them are new) and indicates relationships among them (if possible). We actually present more categories than that are used later, so as to make this chapter a useful reference for categories of stable domains. These categories have been discovered with different motivations. Due to the lack of space we can only explain very briefly the motivations. However, we will try to provide enough information about the categories themselves, and leave it to interested readers to search into the relevant literatures listed in the bibliography for a complete account of applications of stable theories.

This chapter is organized as follows. Section 6.1 presents cartesian closed categories of stable domains. Section 6.2 gives an introduction to categories for concurrency. Section 6.3 introduces monoidal closed categories. In the last section relationships among the categories are discussed.

There are two prominent categories of coherent spaces:

$\mathbf{COH}_s$	—	coherent families with stable functions,
$\mathbf{COH}_l$	—	coherent families with linear functions.

Other categories presented here include:

DI	—	dI-domains with stable functions,
DL	—	dI-domains with linear functions,
$\mathbf{SEV}_s$	—	stable event structures with stable maps,
$\mathbf{SEV}_l$	—	stable event structures with linear maps,
$\mathbf{SEV}^*_{syn}$	—	stable event structures with partially synchronous morphisms,
$\mathbf{SEV}_{syn}$	—	stable event structures with synchronous morphisms,
$\mathbf{SF}_s$	—	stable families with stable functions,
$\mathbf{SF}_l$	—	stable families with linear functions.

Here **DI** is due to Berry [Be78]; $\mathbf{SEV}_s$, $\mathbf{SEV}^*_{syn}$, $\mathbf{SEV}_{syn}$ and $\mathbf{SF}_s$ are due to Winskel [Wi82], [Wi88]; $\mathbf{COH}_s$ and $\mathbf{COH}_l$ are due to Girard [Gi87a], [Gi87b], though presented differently here; and $\mathbf{SEV}_l$ is new (see [Zh91] for more detail). The remaining **DL** and $\mathbf{SF}_l$ are known to exist, but they have not been put into use. Stable domains can also be represented as a special kind of information systems called *prime information systems*. Because of the importance of prime information systems to the logical framework, we will introduce them in a separate chapter (Chapter 7).

6.1 Cartesian Closed Categories

This section presents cartesian closed categories

$$\mathbf{DI},\ \mathbf{SEV}_s,\ \mathbf{SF}_s,\ \text{and}\ \mathbf{COH}_s$$

among the list given above.

6.1.1 DI

DI-domains, or stable domains as some people call, were discovered by Berry from the study of the full-abstraction problem for typed λ-calculi.

They are special kinds of Scott domains which have a more operational nature. The functions between dI-domains are stable functions under an order which takes into account the manner in which they compute. DI-domains with stable functions form a cartesian closed category **DI** [Be78]. This makes them a nice alternative framework in which to do denotational semantics.

Definition 6.1 *A dI-domain is a consistently complete, ω-algebraic cpo D which satisfies*

- *Axiom d:* $\forall x, y, z \in D.\ y \uparrow z \Longrightarrow x \sqcap (y \sqcup z) = (x \sqcap y) \sqcup (x \sqcap z)$, *and*
- *Axiom I:* $\forall d \in D^0.\ |\{x \mid x \sqsubseteq d\}| < \infty$.

Axiom d expresses a distributive property. This property can be violated by many Scott domains, as shown by Example 6.1.

Example 6.1 *Two Scott domains lacking distributivity(axiom d):*

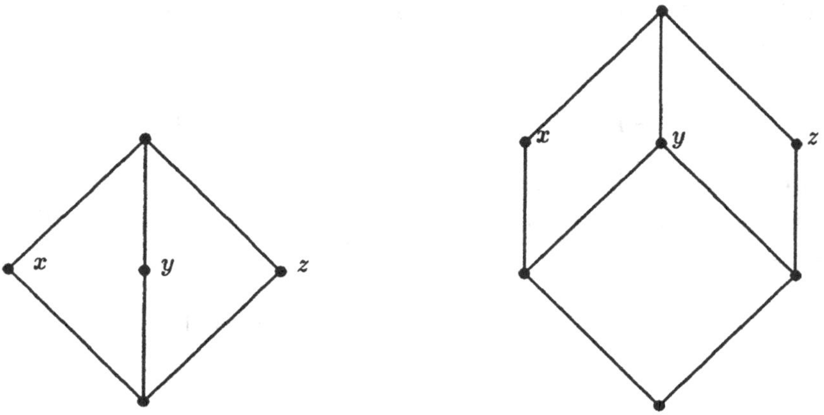

Axiom I says that in a dI-domain D any finite element dominates only finitely many elements. Girard [Gi87b] argues that this is a desirable property for domains, since in this way 'finiteness is finiteness'. There are, of course, Scott domains which do not have this property.

Example 6.2 *A Scott domain which violates axiom I:*

Taking another look at Axiom d, one wonders if the compatibility of y and z there can be replaced by the compatibility of $\{x, y, z\}$. It is indeed possible. The following proposition confirms that Axiom d can be replaced by a seemingly weaker one, requiring distributivity to hold only for compatible triples.

Proposition 6.1 *A domain D satisfies Axiom d if and only if for all $x, y, z \in D$*

$$\{\, x, y, z\,\} \uparrow \Longrightarrow x \sqcap (y \sqcup z) = (x \sqcap y) \sqcup (x \sqcap z).$$

Proof The 'only if' part is trivial.

If: For any compatible pair y, z in D, $\{x \sqcap (y \sqcup z),\ y,\ z\}$ is a compatible set. Therefore,

$$[x \sqcap (y \sqcup z)] \sqcap (y \sqcup z) = \Big[[x \sqcap (y \sqcup z)] \sqcap y\Big] \sqcup \Big[[x \sqcap (y \sqcup z)] \sqcap z\Big],$$

or

$$x \sqcap (y \sqcup z) = (x \sqcap y) \sqcup (x \sqcap z).$$

□

DI-domains are the objects of the category **DI**. Now we present the morphisms of the category: the stable functions.

Definition 6.2 *A function $f : D \to E$ between two dI-domains D and E is stable if it is continuous and preserves meets of pairs of compatible* *elements, i.e.,*

$$\forall x,\ y \in D.\ x \uparrow y \Longrightarrow f(x \sqcap y) = f(x) \sqcap f(y).$$

It is linear if it is stable, and further

$$\forall X \subseteq D.\ X \uparrow \Longrightarrow f(\bigsqcup X) = \bigsqcup \{ f(x) \mid x \in X \}.$$

Let f, g be stable functions from D to E. f is stably less than g, written $f \sqsubseteq_s g$, if

$$\forall x,\ y \in D.\ x \sqsubseteq y \Longrightarrow f(x) = f(y) \sqcap g(x).$$

Write $[D \rightarrow_s E]$ for the set of stable functions from D to E.

DI-domains can be used as a model for typed lambda-calculi, which is sometimes called the stable model. What makes this possible is that **DI** is a cartesian closed category.

Theorem 6.1 *(Berry)* **DI** *is a cartesian closed category. The products are the cartesian product ordered coordinatewise, and the function space of dI-domains D and E consists of the stable functions between them, ordered under the stable order.*

We do not provide a proof of the theorem here directly. Instead, we try to introduce important properties about dI-domains and stable functions to the extent that if the reader wishes to, he can produce such a proof easily.

For convenience, from now on we use p, q for complete primes and a, b for isolated elements.

Before going further into the topic, we introduce a key technique in dealing with dI-domains. It has to do with the relation between distributivity and prime algebraicness. Recall that a complete prime of a cpo D is an element $p \in D$ such that

$$p \sqsubseteq \bigsqcup X \Longrightarrow \exists x \in X . p \sqsubseteq x$$

for all compatible set $X \subseteq D$.

Definition 6.3 *Let D be a consistently complete partial order. D is prime algebraic if*

$$x = \bigsqcup \{ p \mid p \sqsubseteq x \ \&\ \ p \textit{ is a complete prime} \}$$

for all $x \in D$.

There is an important result which states that for a consistently complete partial order D satisfying Axiom I, D is a prime algebraic domain if and only if it is a dI-domain. The proof we are going to offer for this result is different from the original one given by Winskel. Our proof is based on Lemma 6.1, which I have not found in the literature.

To begin with, note that if a domain D satisfies Axiom d then by repeatedly applying Axiom d we have

$$x \sqcap \bigsqcup X = \bigsqcup \{ x \sqcap a \mid a \in X \}$$

for any finite compatible subset $X \subseteq D$. Also note that if X is a compatible subset of a domain D and $X = X_1 \cup X_2$, then we have

$$\bigsqcup X = (\bigsqcup X_1) \sqcup (\bigsqcup X_2).$$

Lemma 6.1 *The complete primes of a dI-domain D are those isolated elements which have a unique element immediately below them. We say x is immediately below y, or x is covered by y, if $x \sqsubseteq y$ and for every z such that $x \sqsubseteq z \sqsubseteq y$, either $x = z$ or $y = z$.*

Proof Let p be a complete prime and let X be the finite (by Axiom I) set of isolated elements immediately below it. X must be non-empty since $\perp$ is not a complete prime. But X cannot have more than one element, for otherwise we have $p \sqsubseteq \bigsqcup X$ but not $p \sqsubseteq x$ for any $x \in X$. Therefore there is a unique element immediately below p.

Suppose, on the other hand, there is a unique element w immediately below an isolated element $a \in D$. Let X be a compatible set such that

$$a \sqsubseteq \bigsqcup X.$$

We can assume that X is finite since a is isolated. Therefore

$$a = a \sqcap (\bigsqcup X) = \bigsqcup \{ a \sqcap x \mid x \in X \}.$$

Clearly $a \sqcap x \sqsubseteq a$. If $a \sqcap x \neq a$ then $a \sqcap x \sqsubseteq w$, since w is the unique element immediately below a. Therefore for $a \sqsubseteq \bigsqcup X$ to hold, we must have $a \sqcap x = a$ for some $x \in X$. That means, $a \sqsubseteq x$ for some $x \in X$. Hence a is a complete prime.

□

Theorem 6.2 *(Winskel) Suppose D is a Scott domain satisfying Axiom I. Then D is prime algebraic if and only if it is a dI-domain.*

Proof Assume D is a prime algebraic Scott domain satisfying Axiom I. Let $x \uparrow y$ and

$$x = \bigsqcup\{ p \sqsubseteq x \mid p \in D^p \},$$
$$y = \bigsqcup\{ q \sqsubseteq y \mid q \in D^p \}.$$

Given that p and q's are complete primes, we get

$$x \sqcup y = \bigsqcup(\{ p \sqsubseteq x \mid p \in D^p \} \cup \{ q \sqsubseteq y \mid q \in D^p \}),$$
$$x \sqcap y = \bigsqcup(\{ p \sqsubseteq x \mid p \in D^p \} \cap \{ q \sqsubseteq y \mid q \in D^p \}).$$

Axiom d then follows from the prime algebraicness of D.

Assume D is a dI-domain, on the other hand. We show that it is prime algebraic. It is enough to show that

$$\forall d \in D^0 . d = \bigsqcup\{ p \sqsubseteq d \mid p \in D^p \},$$

by the algebraicness of D. If $d \in D^0$ is already a complete prime, there is nothing to prove. If $d \in D^0$ is not a complete prime then

$$d = \bigsqcup\{ a \sqsubseteq d \mid a \text{ is covered by } d \},$$

by Lemma 6.1. For each a covered by d, it is either a complete prime or an isolated element but not a complete prime. For the latter case a is again the least upper bound of those elements covered by a. Reasoning in this way repeatedly, we can finally get down to d_1's each having a unique element below it. Those d_1's, by Lemma 6.1, are complete primes. Going upwards from those d_1's we see that the original d is the least upper bound of the complete primes below it.

□

The full-abstraction problem for typed lambda-calculi lead Berry to consider the problem of capturing a notion of 'sequential functions'. As one of the possible candidates for sequential functions Berry introduced stable functions so that non-sequential functions like 'parallel-or' (see Section 8.1) are excluded. Stable functions have a property that their values are totally determined by those at some minimal points. To describe the minimal points we introduce the following definition.

Definition 6.4 *Let $f : D \to E$ be a stable function from a dI-domain D to a dI-domain E. Define μf to be the set of pairs such that $(a, p) \in \mu f$ if*

$$f(a) \sqsupseteq p \ \& \ [\forall a' \sqsubseteq a.\ f(a') \sqsupseteq p \Longrightarrow a = a'],$$

where $a \in D^0$, the set of finite elements of D and $p \in E^p$, the set of complete primes of E.

One can then understand a pair $(a, p) \in \mu f$ as saying that a is a element point for f to assume value p.

The following two lemmas are useful. From the first lemma we know that f can be recovered from the set μf.

Lemma 6.2 *Suppose $f : D \to E$ is a stable function. Then for any $x \in D$,*

$$f(x) = \bigsqcup\{\, p \mid \exists a \sqsubseteq x.\ (a,\ p) \in \mu f \,\}.$$

Proof Let $f : D \to E$ be a stable function. We have $f(a) \sqsupseteq p$ for any $(a, p) \in \mu f$. Therefore if $x \sqsupseteq a$ and $(a, p) \in \mu f$, then $f(x) \sqsupseteq p$. Hence

$$f(x) \sqsupseteq \bigsqcup\{\, p \mid \exists a \sqsubseteq x.\ (a,\ p) \in \mu f \,\}.$$

On the other hand, by Axiom I we know that for any complete prime q in E such that $q \sqsubseteq f(x)$, there is an element $b \sqsubseteq x$ for which $(b, q) \in \mu f$. This means

$$q \sqsubseteq \bigsqcup\{\, p \mid \exists a \sqsubseteq x.\ (a,\ p) \in \mu f \,\}.$$

But E is a dI-domain, hence prime algebraic (Theorem 6.2). Therefore $f(x) = \bigsqcup\{\, q \mid q \in E^p \ \&\ q \sqsubseteq f(x) \,\} \sqsubseteq \bigsqcup\{\, p \mid \exists a \sqsubseteq x.\ (a,\ p) \in \mu f \,\}$.

□

The second lemma implies that compatible stable functions have the same minimal point related to a given value they can all assume.

Lemma 6.3 *Let $f, g : D \to E$ be stable functions. We have*

1. *If $f \sqsubseteq_s g$ then $\mu f \subseteq \mu g$,*
2. *$[(a,\ p),\ (a',\ p) \in \mu f \ \&\ a \uparrow a'] \Longrightarrow a = a'$.*

Proof Suppose $f, g : D \to E$ are stable functions such that $f \sqsubseteq_s g$. For any $(a, p) \in \mu f$, $p \sqsubseteq f(a) \sqsubseteq g(a)$. Let

$$b = \bigsqcap\{\, x \mid x \sqsubseteq a \ \&\ g(x) \sqsupseteq p \,\}.$$

Apparently $b \sqsubseteq a$, and, by Axiom I, $(b, p) \in \mu g$. $f \sqsubseteq_s g$ implies, by definition,

$$f(b) = f(a) \sqcap g(b).$$

Hence $f(b) \sqsupseteq p$, as $f(a) \sqsupseteq p$ and $g(b) \sqsupseteq p$ (using Axiom I and stability). We must have $a = b$, because $(a, p) \in \mu f$. This means $(a, p) \in \mu g$, too.

Suppose f is stable, $a \uparrow a'$, and (a, p), (a', p) are members of μf. It follows that $f(a) \sqsupseteq p$, $f(a') \sqsupseteq p$. As $a \uparrow a'$, we have

$$f(a \sqcap a') = f(a) \sqcap f(a') \sqsupseteq p.$$

Thus $a \sqcap a' = a$, $a \sqcap a' = a'$. Hence $a = a'$.

□

There is an easy to prove, but useful fact about stable functions.

Lemma 6.4 *If f and g are compatible stable functions and x, y are compatible, then*

$$\begin{aligned} f(x) \sqcap g(y) &= f(x \sqcap y) \sqcap g(x \sqcap y) \\ &= f(y) \sqcap g(x). \end{aligned}$$

Suppose we have a compatible set of stable functions form D to E. How can we determine the least upper bound and greatest lower bound (with respect to the stable order, of course) of the set? The next two lemmas show that they can be derived pointwise.

Lemma 6.5 *If $F \subseteq [D \to_s E]$ is compatible (with respect to the stable order) then*

$$\bigsqcup F = \lambda x. \bigsqcup_{f \in F} f(x).$$

Proof It is enough to show that $\lambda x. \bigsqcup_{f \in F} f(x)$ is continuous, stable and greater than any function in F under the stable order.

Continuity: Let

$$x_0 \sqsubseteq x_1 \sqsubseteq \cdots \sqsubseteq x_n \sqsubseteq \cdots$$

be a chain in D. Obviously

$$\begin{aligned} \bigsqcup_{f \in F} f\Big(\bigsqcup_{i \in \omega} x_i\Big) &= \bigsqcup_{f \in F} \Big(\bigsqcup_{i \in \omega} f(x_i)\Big) \\ &= \bigsqcup_{i \in \omega} \Big(\bigsqcup_{f \in F} f(x_i)\Big). \end{aligned}$$

Hence it is continuous.

Stability: For any compatible pair x, y in D, we have, by virtue of the prime algebraic property of D and Lemma 6.4,

$$\begin{aligned}\bigsqcup_{f\in F} f(x\sqcap y) &\sqsubseteq \Big(\bigsqcup_{f\in F} f(x)\Big)\sqcap\Big(\bigsqcup_{g\in F} g(y)\Big)\\ &= \bigsqcup_{f\in F}\bigsqcup_{g\in F}\Big(f(x)\sqcap g(y)\Big)\\ &= \bigsqcup_{f\in F}\bigsqcup_{g\in F}\Big(f(x\sqcap y)\sqcap g(x\sqcap y)\Big)\\ &\sqsubseteq \bigsqcup_{f\in F} f(x\sqcap y).\end{aligned}$$

Therefore

$$\bigsqcup_{f\in F} f(x\sqcap y) = \Big(\bigsqcup_{f\in F} f(x)\Big)\sqcap\Big(\bigsqcup_{g\in F} g(y)\Big).$$

Let $y \sqsubseteq z$ and $g \in F$. We have

$$\begin{aligned}g(y) &\sqsubseteq g(z)\sqcap\bigsqcup_{f\in F} f(y)\\ &= \bigsqcup_{f\in F}\Big(g(z)\sqcap f(y)\Big)\\ &= \bigsqcup_{f\in F}\Big(g(y\sqcap z)\sqcap f(y\sqcap z)\Big)\\ &\sqsubseteq g(y),\end{aligned}$$

which implies

$$g(y) = g(z)\sqcap\bigsqcup_{f\in F} f(y).$$

In other words, $g \sqsubseteq_s \lambda x.\bigsqcup_{f\in F} f(x)$.

□

In fact, Lemma 6.5 confirms the consistent completeness of the stable function space $D \to_s E$. To show that $D \to_s E$ is indeed a dI-domain again, we need to check Axiom d and Axiom I. Axiom d follows from Lemma 6.5 and the following Lemma 6.6. Axiom I will be dealt with later in Theorem 6.3.

Lemma 6.6 *If $F \subseteq [D \to_s E]$ is compatible then*

$$\bigsqcap F = \lambda x.\bigsqcap_{f\in F} f(x).$$

Proof First we verify the non-trivial fact that $\lambda x. \bigsqcap_{f \in F} f(x)$ is continuous. Let

$$x_0 \sqsubseteq x_1 \sqsubseteq \cdots \sqsubseteq x_n \sqsubseteq \cdots$$

be a chain in D. What is needed is the equation

$$\bigsqcap_{f \in F} \Big[\bigsqcup_{i \in \omega} f(x_i)\Big] = \bigsqcup_{i \in \omega}\Big[\bigsqcap_{f \in F} f(x_i)\Big].$$

Clearly

$$\bigsqcap_{f \in F} \Big[\bigsqcup_{i \in \omega} f(x_i)\Big] \sqsupseteq \bigsqcup_{i \in \omega}\Big[\bigsqcap_{f \in F} f(x_i)\Big].$$

The argument for the inequality in the other direction runs as follows:

$$\begin{aligned} d \in D^0 \mathbin{\&} \bigsqcap_{f \in F} \Big[\bigsqcup_{i \in \omega} f(x_i)\Big] \sqsupseteq d & \\ \Longrightarrow \forall f \in F. \bigsqcup_{i \in \omega} f(x_i) \sqsupseteq d & \\ \Longrightarrow \forall f \in F\, \exists i_f.\, f(x_{i_f}) \sqsupseteq d & \\ \Longrightarrow \bigsqcap_{f \in F} f(x_{i_f}) \sqsupseteq d. & \end{aligned}$$

Let $i_h = \min \{ i_f \mid f \in F \}$ where $h \in F$. We have, by Lemma 6.4,

$$\forall f \in F.\ h(x_{i_h}) \sqcap f(x_{i_f}) = h(x_{i_h}) \sqcap f(x_{i_h}).$$

Therefore,

$$\bigsqcap_{f \in F} f(x_{i_f}) = \bigsqcap_{f \in F} f(x_{i_h}),$$

which implies

$$\bigsqcup_{i \in \omega}\Big[\bigsqcap_{f \in F} f(x_i)\Big] \sqsupseteq \bigsqcap_{f \in F} f(x_{i_h}) \sqsupseteq d.$$

That $\lambda x. \bigsqcap_{f \in F} f(x)$ is stable and $\lambda x. \bigsqcap_{f \in F} f(x) \sqsubseteq_s g$ for all $g \in F$ are routine.

□

Notice that we require F to be compatible in Lemma 6.6. Greatest lower bounds always exist, but may not equal to the function specified pointwise.

Now we are in a position to explain why Axiom I is necessary from another angle. It has to do with the fact that stable functions between ω-algebraic, distributive Scott domains under the stable order need not form an ω-algebraic domain: There can be too many finite elements in the stable function space. Consider the stable function space from the domain given in Example 6.2 to itself, where the order of elements in the domain is the reverse of the order for natural numbers. Each stable function f corresponds to an infinite sequence

$$(f(1), f(2), f(3), \cdots, f(n), \cdots).$$

We claim that *all* such stable functions except the bottom are complete primes of the stable function space!

Let $\{\phi_i \mid i \in I\}$ be a compatible set of stable functions such that

$$\bigsqcup_{i \in I} \phi_i \sqsupseteq_s f,$$

where f is not the bottom. By Lemma 6.5 we have, in particular,

$$\bigsqcup_{i \in I} \phi_i(1) \sqsupseteq f(1).$$

Therefore, there exists $i_0 \in I$ such that $\phi_{i_0}(1) \sqsupseteq f(1)$, as $f(1)$ is a complete prime (note $f(1) \neq \bot$). Since ϕ_{i_0} and f are compatible, we have, by Lemma 6.4, for all $k \geq 1$, $f(k) \sqcap \phi_{i_0}(1) = \phi_{i_0}(k) \sqcap f(1)$. Hence for all $k \geq 1$,

$$\begin{aligned} f(k) &= f(k) \sqcap f(1) \\ &= f(k) \sqcap \phi_{i_0}(1) \\ &= \phi_{i_0}(k) \sqcap f(1) \\ &\sqsubseteq \phi_{i_0}(k). \end{aligned}$$

This implies $f \sqsubseteq_s \phi_{i_0}$, which shows that f is a complete prime in the stable function space. By the diagonal argument we deduce that there are uncountably many such functions. Thus the stable function space is no longer ω-algebraic.

To show that $D \rightarrow_s E$ entertains Axiom I, we now deal with the structure of stable functions. More precisely, we introduce a characterization of finite elements in the stable function space. It is a generalization of the notion of traces proposed by Girard [Gi87b] from coherent spaces to stable domains.

Theorem 6.3 *Let $f, g \in [D \rightarrow_s E]$. $f \sqsubseteq_s g$ if and only if $\mu f \subseteq \mu g$. For $\{ (a_i, p_i) \mid i \in I \} \subseteq D^0 \times E^p$,*

$$\{ (a_i, p_i) \mid i \in I \} = \mu f$$

for some $f \in [D \rightarrow_s E]$ *if and only if*

- $\forall J \Subset I. \{ a_i \mid i \in J \} \uparrow \Longrightarrow \{ p_i \mid i \in J \} \uparrow$,
- $[a_i \uparrow a_j \ \& \ (p_i = p_j)] \Longrightarrow (a_i = a_j)$, *and*
- $\forall p \in E^p. [p_i \sqsupseteq p \Longrightarrow \exists j. p_j = p \ \& \ a_i \sqsupseteq a_j]$.

Proof Lemma 6.3 shows that $f \sqsubseteq_s g$ implies $\mu f \subseteq \mu g$. Suppose, on the other hand, that $\mu f \subseteq \mu g$. It follows from Lemma 6.2 that

$$\forall x \in D.\ f(x) \sqsubseteq g(x).$$

To prove $f \sqsubseteq_s g$ we have to show that for x, y in D, $x \sqsubseteq y$ implies

$$f(x) = f(y) \sqcap g(x).$$

To this end let $p \in E^p$ and $p \sqsubseteq f(y) \sqcap g(x)$. Clearly there exist $a \sqsubseteq x$ and $b \sqsubseteq y$ such that $(b,\ p) \in \mu f$ and $(a,\ p) \in \mu g$. However $\mu f \subseteq \mu g$; we have $(b,\ p) \in \mu g$. By Lemma 6.3, $a = b$. This implies that $f(x) \sqsupseteq f(b) \sqsupseteq p$. By the prime algebraicness of E we get

$$f(x) \sqsupseteq f(y) \sqcap g(x),$$

enough for the equation

$$f(x) = f(y) \sqcap g(x)$$

to hold.

To prove the second part, suppose $\{ (a_i,\ p_i) \mid i \in I \} = \mu f$ for some stable function f. It is routine to check that the three properties mentioned in Theorem 6.3 hold.

Let $\{ (a_i,\ p_i) \mid i \in I \} \subseteq D^0 \times E^p$ be a set with the three properties. We show that the stable function f for which (Lemma 6.2 concludes that such a function is unique)

$$\{ (a_i, p_i) \mid i \in I \} = \mu f$$

can be obtained as pointwise least upper bounds $\bigsqcup_{i \in I} [a_i, p_i]$ where

$$[a,p](x) = \begin{cases} p & \text{if } x \sqsupseteq a\ , \\ \bot & \text{otherwise.} \end{cases}$$

Obviously $\bigsqcup_{i\in I}[a_i, p_i]$ is continuous. To check stability let $x, y \in D$ and $x \uparrow y$. Suppose

$$p \sqsubseteq \bigsqcup_{i\in I}[a_i, p_i](x) \sqcap \bigsqcup_{i\in I}[a_i, p_i](y)$$

where $p \in E$ is a complete prime. We have, for some i, j, $p \sqsubseteq p_i$, $a_i \sqsubseteq x$ and $p \sqsubseteq p_j$, $a_j \sqsubseteq y$. By the third property, there exist s, t such that $p_s = p$, $a_s \sqsubseteq a_i$ and $p_t = p$, $a_t \sqsubseteq a_j$. Therefore $a_s = a_t$ as $a_s \uparrow a_t$ and $p_s = p_t$. We now have $a_s = a_t \sqsubseteq x \sqcap y$ and

$$p \sqsubseteq \bigsqcup_{i\in I}[a_i, p_i](x \sqcap y).$$

Since E is prime algebraic,

$$\bigsqcup_{i\in I}[a_i, p_i](x \sqcap y) \sqsupseteq \bigsqcup_{i\in I}[a_i, p_i](x) \sqcap \bigsqcup_{i\in I}[a_i, p_i](y).$$

This implies that $\bigsqcup_{i\in I}[a_i, p_i]$ is stable. It remains to show that

$$\{\, (a_i,\, p_i) \mid i \in I \,\} = \mu f$$

where we abbreviate $\bigsqcup_{i\in I}[a_i, p_i]$ as f. We have

$$\begin{aligned} f(a_j) &= \bigsqcup\{\, p_i \mid a_i \sqsubseteq a_j \,\} \\ &\sqsupseteq p_j. \end{aligned}$$

Let $y \sqsubseteq a_j$ and $f(y) \sqsupseteq p_j$, *i.e.*, $\bigsqcup\{\, p_i \mid a_i \sqsubseteq y \,\} \sqsupseteq p_j$. Since p_j is a complete prime, $p_i \sqsupseteq p_j$ for some i with $a_i \sqsubseteq y$. The third condition of the theorem implies the existence of some k such that $p_k = p_j$ and $a_k \sqsubseteq a_i$. But $a_k = a_j$ since $a_k \uparrow a_j$. Hence $y = a_j$. This means $(a_j,\, p_j) \in \mu f$.

For any $(a,\, p) \in \mu f$, we have $f(a) \sqsupseteq p$. Therefore

$$\bigsqcup\{\, p_i \mid a_i \sqsubseteq a \,\} \sqsupseteq p.$$

Since p is a complete prime, there is some p_i such that $p_i \sqsupseteq p$. By the third condition again, $p_j = p$ for some j such that $a_j \sqsubseteq a_i$. By the result from the previous paragraph we have $(a_j,\, p_j) \in \mu f$. Therefore $a_j = a$ by Lemma 6.3 (taking $f = g$).

□

We call a finite set of pairs with the three properties given in Theorem 6.3 *stable joinable.*

6.1.2 $\mathbf{SEV}_s$

Event structures are models for processes of concurrent computation. An event structure is a description of a set of events in terms of consistency and enabling relations. The consistency relation indicates whether some events can occur together or not, and the enabling relation specifies the condition when a particular event may occur with regards to the occurrence of other events.

Definition 6.5 *An event structure is a triple*

$$\underline{E} = (E,\ Con,\ \vdash)$$

where

- E *is a countable set of events,*
- $Con \subseteq \{B \mid B \Subset E\}$ *is the consistency predicate which satisfies*
$$X \subseteq Y \;\&\; Y \in Con \Longrightarrow X \in Con,$$
- $\vdash \subseteq Con \times E$ *is the enabling relation which satisfies*
$$(X \vdash e \;\&\; X \subseteq Y \;\&\; Y \in Con) \Longrightarrow Y \vdash e.$$

When we have the enabling relation $X \vdash e$ in an event structure, we say that the event e is enabled by the set of events X. We can also say that the occurrences of all events in X is sufficient for e to occur. Note following this intuition, $\{e\} \vdash e$ does not contribute any information to an event structure.

A configuration of an event structure is a set of events which is consistent, and each of its event is enabled by a set of events of the configuration occurred previously. That is, a configuration is a set of events which have occurred by certain stage in a process.

Definition 6.6 *Let* $\underline{E} = (E,\ Con,\ \vdash)$ *be an event structure. A configuration of* $\underline{E}$ *is a subset* $x \subseteq E$ *which is*

- *consistent:* $X \Subset x \Longrightarrow X \in Con,$
- *secured:* $\forall e \in x\ \exists e_0, e_1, \cdots, e_n \in x.\ e_n = e\ \&$
$$\forall i \leq n.\ \{e_k \mid 0 \leq k \leq i-1\} \vdash e_i.$$

Write the set of configurations of an event structure $\underline{E}$ *as* $\mathcal{F}(\underline{E})$.

There is a special class of event structures for which each configuration determines a partial order of causal dependency on the events. Intuitively, an event e_1 causally depends on an event e_0 if the occurrence of the event e_0 is necessary in order for the event e_1 to occur. Event structures of this kind are called *stable*.

Definition 6.7 *An event structure $\underline{E}$ is stable if it satisfies the following axiom*

$$(X \vdash e \;\&\; Y \vdash e \;\&\; X \cup Y \cup \{e\} \in Con) \Longrightarrow X \cap Y \vdash e.$$

We mentioned that a stable event structure is an event structure each configuration of which determines a partial order of causal dependency on the events of the configuration. This partial order can now be specified as follows.

Definition 6.8 *Let $\mathcal{F}(\underline{E})$ be the family of configurations of a stable event structure $\underline{E}$. Let $x \in \mathcal{F}(\underline{E})$ and $e, e' \in x$. Define*

$$e' \leq_x e \text{ if } \forall y \in \mathcal{F}(\underline{E}).\,(e \in y \;\&\; y \subseteq x \Longrightarrow e' \in y),$$
$$\lceil e \rceil_x = \bigcap \{ y \in \mathcal{F}(\underline{E}) \mid e \in y \;\&\; y \subseteq x \}.$$

Note $\leq_x$ is a partial order and $\lceil e \rceil_x = \{ e' \in x \mid e' \leq_x e \}$. Given a stable event structure $\underline{E}$, the complete primes in $\mathcal{F}(\underline{E})$ are of the form $\lceil e \rceil_x$.

A *prime event structure* is a stable event structure such that the enabling relation is a finitary partial order (*i.e.* one that satisfies Axiom I) on events and the consistency relation *Con* on the set of finite subsets of events satisfying the following:

$$Y \subseteq X \;\&\; X \in Con \Longrightarrow Y \in Con,$$
$$X \in Con \;\&\; e \in X \;\&\; e \leq e' \Longrightarrow X \cup \{e\} \in Con.$$

For a prime event structure, the complete primes are uniquely determined by the events. Hence they can be written as $\lceil e \rceil$ without referring to any configuration x.

Theorem 6.4 *(Winskel) Let $\underline{E}$ be a stable event structure. Then*

$$(\mathcal{F}(\underline{E}), \subseteq)$$

is a dI-domain.

The proof of this theorem goes by first showing that $(\mathcal{F}(\underline{E}), \subseteq)$ is a prime algebraic Scott domain satisfying Axiom I, and then applying Theorem 6.2. To find a decent definition of morphisms on stable event structures (stable functions were used as morphisms, which was a bit unsatisfactory since they are the morphisms for the category **DI**) note that there is a useful notion of minimal entailment. Let $\underline{E} = (E, Con, \vdash)$ be a stable event structure. Suppose $Y \vdash e$. Then we have the minimal entailment $X \vdash_{min} e$, where

$$X = \bigcap \{X' \mid X' \subseteq Y \ \&\ X' \vdash e\}.$$

Suppose $X \vdash_{min} e$. Then $X \vdash e$. Furthermore, for any Z, $Z = X$ if $X \cup Z \in Con$ and $Z \vdash_{min} e$. Of course interesting minimal entailments are those relations $X \vdash_{min} e$ with $e \notin X$.

Definition 6.9 *Let* $\underline{E} = (E, Con_{\underline{E}}, \vdash_{\underline{E}})$, $\underline{F} = (F, Con_{\underline{F}}, \vdash_{\underline{F}})$ *be stable event structures. A stable map from* $\underline{E}$ *to* $\underline{F}$ *is a relation* $R \subseteq \Phi \times F$ *which satisfies*

1. *Compatibility*: $T \Subset R \ \&\ \bigcup(\pi_1 T) \in Con_{\underline{E}} \Longrightarrow \pi_2 T \in Con_{\underline{F}}$,
2. *Minimality*: $(x \cup y \in \mathcal{F}(\underline{E}) \ \&\ x\, R\, e \ \&\ y\, R\, e) \Longrightarrow x = y$, *and*
3. *Completeness*: $x\, R\, e \Longrightarrow \exists X.(X \vdash_{min} e \ \&\ \forall e' \in X \exists x' \subseteq x.\ x'\, R\, e')$.

Here Φ *stands for the set of finite elements of* $\mathcal{F}(\underline{E})$.

Stable event structures with stable maps form a category. The identity Id is given by $\lceil e \rceil_x\, Id\, e$ if $e \in x$ and $x \in \mathcal{F}(E)$. To compose stable maps, define, for a stable map R, $\tilde{R}$ to be a relation such that $a\, R\, b$ if and only if there exist a_i, p_i such that $a_i\, R\, p_i$, $1 \leq i \leq n$, $a = \bigcup_{1 \leq i \leq n} a_i \in \mathcal{F}(\underline{E})$, and $b = \{p_i \mid 1 \leq i \leq n\} \in \mathcal{F}(\underline{F})$. It is not difficult to check that $\tilde{R}$ has the following properties:

1. $H \Subset \tilde{R} \ \&\ \bigcup(\pi_1 H) \in Con_{\underline{E}} \Longrightarrow \pi_2 H \in Con_{\underline{F}}$,
2. $(a \cup b \in \mathcal{F}(\underline{E}) \ \&\ a\, \tilde{R}\, c \ \&\ b\, \tilde{R}\, c) \Longrightarrow a = b$,
3. $a\, \tilde{R}\, b \supseteq c \Longrightarrow \exists a' \subseteq a.\ a'\, \tilde{R}\, c$.

Now we can compose stable maps $R : \underline{E} \to \underline{F}$ and $S : \underline{F} \to \underline{G}$ in the following way. Define $R \circ S$ to be a relation on $P_{\underline{E}} \times C$ by letting

$$a\,(R \circ S)\,p \Longleftrightarrow \exists b.\ a\, \tilde{R}\, b \ \&\ b\, S\, p.$$

From the properties of $\tilde{R}$ one can see that $R \circ S : \underline{E} \to \underline{G}$ is indeed a stable map.

The construction of function space of stable event structures is given as follows.

Definition 6.10 *Let* $\underline{E_1} = (E_1, Con_1, \vdash_1)$ *and* $\underline{E_2} = (E_2, Con_1, \vdash_2)$ *be stable event structures. Their function space,* $[\underline{E_1} \to \underline{E_2}]$*, is the structure* $(E, Con, \vdash,)$ *where*

$$\begin{aligned}
&E = \{(x, e_2) \mid x \in \mathcal{F}(\underline{E_1}),\, e_2 \in E_2\},\\
&X \in Con \;\; \text{if} \;\; \forall Y \subseteq X.\, \pi_1 Y \in Con_1 \Longrightarrow \pi_2 Y \in Con_2 \;\&\\
&\qquad\qquad \forall a, b \in X.(\pi_1 a \cup \pi_1 b \in \mathcal{F}(\underline{E_1}) \;\&\; \pi_2 a = \pi_2 b) \Longrightarrow a = b,\\
&X \vdash (x, e_2) \;\; \text{if} \;\; \{e \mid (y, e) \in X \;\&\; y \subseteq x\} \vdash_2 e_2.
\end{aligned}$$

We have

Theorem 6.5 *The function space of two stable event structures is a stable event structure. R is a configuration of* $[\underline{E_1} \to \underline{E_2}]$ *if and only if*

$$R : \underline{E_1} \to \underline{E_2}$$

is a stable map. $R : \underline{E_1} \to \underline{E_2}$ *is a stable map if and only if*

$$f(x) = \{e \in E_2 \mid \exists x' \subseteq x.\, (x', e) \in R\}$$

specifies a stable function from $\mathcal{F}(\underline{E_1})$ *to* $\mathcal{F}(\underline{E_2})$*. The stable order is captured by the subset relation on the stable maps.*

As a more general result, we have

Theorem 6.6 DI *is equivalent to* $\mathbf{SEV_s}$.

The equivalence of categories is in the precise sense of [Ma71]. However, by one of the results in [Ma71] it is enough to show that there exists a functor from $\mathbf{SEV_s}$ to **DI** which is full and faithful, and each dI-domain D is isomorphic to one that is determined by some stable event structure. The material presented so far should contain sufficient information about what the functors should be. It is easy to find a stable event structure $\mathcal{E}(D)$ which determines a dI-domain isomorphic to D. The events of $\mathcal{E}(D)$

are the complete primes of D. The consistency relation is taken to be the compatibility relation, and the entailment is given by $X \vdash e$ if

$$\{p \in D^p \mid p \sqsubseteq e\} \subseteq X.$$

It can be seen that the configurations of this event structure are the unions of downwards closed subsets of D. The desired isomorphism, therefore, exists.

6.1.3 $\mathbf{SF}_s$

Forgetting about the enabling and consistency relation but keeping the configurations of a stable event structure, one gets a *stable family. Stable families* are an axiomatization of the configurations determined by stable event structures.

Definition 6.11 *A stable family is a set of subsets $\mathcal{F}$ of a countable set E which satisfy*

- *finite completeness:*
  $\forall X \subseteq \mathcal{F}.(\forall Y \Subset X.\, Y\uparrow) \Longrightarrow \bigcup X \in \mathcal{F}$,
- *finiteness:*
  $\forall x \in \mathcal{F}\, \forall e \in x \exists z \in \mathcal{F}.\, (\mid z \mid < \infty \;\&\; e \in z \;\& z \subseteq x\,)$,
- *stability:*
  $\forall X \subseteq \mathcal{F}.\, (X \neq \emptyset \;\&\; X\uparrow) \Longrightarrow \bigcap X \in \mathcal{F}$,
- *coincidence freeness:*
  $\forall x \in \mathcal{F}\, \forall e, e' \in x.\, (e \neq e' \Longrightarrow (\exists y \in \mathcal{F}.\, y \subseteq x \;\&\; [e \in y \Leftrightarrow e' \notin y]))$.

Let $\mathcal{F}$ be a finitary family over E. The elements of $\mathcal{F}$ are called configurations and the elements of E events. $\mathcal{F}$ is said to be *full* if $\bigcup \mathcal{F} = E$. If $\underline{E}$ is a stable event structure then $\mathcal{F}(\underline{E})$ is a stable family.

A stable family with set-inclusion gives a dI-domain. Stable families with stable functions as morphisms form a cartesian closed category $\mathbf{SF}_s$. $\mathbf{SF}_s$ is equivalent to $\mathbf{SEV}_s$. Similar to $\mathbf{SEV}_s$, one can define constructions like sum, product, and stable function space in $\mathbf{SF}_s$.

6.1.4 $\mathbf{COH}_s$

Coherent spaces are a special kind of dI-domains, or event structures. As a dI-domain, a coherent space is a domain D which is *coherent*, or *pairwise complete* in the sense that it has least upper bounds for pairwise compatible sets, and p is a complete prime if and only if

$$p \neq \bot_D \;\&\; (x \sqsubset p \Longrightarrow x = \bot_D).$$

As a stable event structure, a coherent space is a structure of the form

$$(\,E,\, Con,\, \{\,\emptyset \vdash e \mid e \in E\,\}\,),$$

where *Con* is determined by a conflict relation.

We present coherent spaces as a special kind of stable families called *coherent families.* It is assumed that coherent families are full as stable families. Thus it is not necessary to specify the event set.

Definition 6.12 *Let $\mathcal{F}$ be a stable family. It is called a coherent family if*

- $\forall x \in \mathcal{F}.\, y \subseteq x \Longrightarrow y \in \mathcal{F}$,
- $(\,X \subseteq \mathcal{F} \;\&\; \forall x,\, y \in X.\, x \uparrow y\,) \Longrightarrow \bigcup X \in \mathcal{F}$.

In fact the above two conditions imply the axiom of finiteness, coincidence-freeness, and stability for stable families. So they alone are enough to determine a coherent family.

Proposition 6.2 *Let $\mathcal{F}$ be a coherent family. Then $(\,\mathcal{F},\, \subseteq\,)$ is a dI-domain with $\{\,\{\,a\,\} \mid \{\,a\,\} \in \mathcal{F}\,\}$ the set of its complete primes.*

Coherent families with stable functions form a cartesian closed category. The related constructions are defined as follows.

Definition 6.13 *(Sum) Let $\mathcal{F}_1$, $\mathcal{F}_2$ be coherent families. Their* sum,

$$\mathcal{F}_1 + \mathcal{F}_2,$$

is the family of subsets which satisfies

$$x \in \mathcal{F}_1 + \mathcal{F}_2 \Longleftrightarrow \begin{array}{l} \bullet\; x \subseteq (\{\,1\,\} \times \bigcup \mathcal{F}_1) \cup (\{\,2\,\} \times \bigcup \mathcal{F}_2) \;\& \\ \bullet\; \exists x_1 \in \mathcal{F}_1.\, x = \{\,1\,\} \times x_1 \text{ or } \exists x_2 \in \mathcal{F}_2.\, x = \{\,2\,\} \times x_2. \end{array}$$

Definition 6.14 *(Product) Let $\mathcal{F}_1$, $\mathcal{F}_2$ be coherent families. Their* product, *$\mathcal{F}_1 \times \mathcal{F}_2$, is the family of subsets which satisfies*

$$x \in \mathcal{F}_1 \times \mathcal{F}_2 \Longleftrightarrow \begin{array}{l} \bullet\; x \subseteq (\{\,1\,\} \times \bigcup \mathcal{F}_1) \cup (\{\,2\,\} \times \bigcup \mathcal{F}_2) \;\& \\ \bullet\; \exists x_1 \in \mathcal{F}_1 \exists x_2 \in \mathcal{F}_2.\, x = \{\,1\,\} \times x_1 \cup \{\,2\,\} \times x_2. \end{array}$$

It is easy to see that the sum and the product of coherent families are still coherent families.

Definition 6.15 *(Stable Function Space) Let $\mathcal{F}_1$, $\mathcal{F}_2$ be coherent families. Their stable function space, $\mathcal{F}_1 \to \mathcal{F}_2$, is the family of subsets which satisfies*

$$x \in \mathcal{F}_1 \to \mathcal{F}_2 \iff \begin{array}{l} \bullet\ x \subseteq \{\, y \in \mathcal{F}_1 \mid |\, y \,| < \infty \,\} \times \bigcup \mathcal{F}_2, \\ \bullet\ (a,\ b \in x \ \&\ \pi_1 a \cup \pi_1 b \in \mathcal{F}_1) \Longrightarrow \{\pi_2 a,\ \pi_2 b\} \in \mathcal{F}_2, \\ \bullet\ [y_1 \cup y_1' \in \mathcal{F}_1 \ \&\ (\, y_1,\ e_2\,),\ (\, y_1',\ e_2\,) \in x] \Longrightarrow y_1 = y_1'. \end{array}$$

Proposition 6.3 *The family $\mathcal{F}_1 \to \mathcal{F}_2$ is coherent and is isomorphic to $[\mathcal{F}_1 \to_s \mathcal{F}_2]$, the dI-domain of stable functions from $\mathcal{F}_1$ to $\mathcal{F}_2$.*

Proof It is routine to show that $\mathcal{F}_1 \to \mathcal{F}_2$ is a coherent family.

The isomorphism between $\mathcal{F}_1 \to \mathcal{F}_2$ and $[\mathcal{F}_1 \to_s \mathcal{F}_2]$ is given by the following pair of functions:

$$\begin{array}{l} Pt : \ (\mathcal{F}_1 \to \mathcal{F}_2) \to [\mathcal{F}_1 \to_s \mathcal{F}_2], \\ \mu : \ [\mathcal{F}_1 \to_s \mathcal{F}_2] \to (\mathcal{F}_1 \to \mathcal{F}_2). \end{array}$$

Here

$$Pt\, x\, (x_1) = \{\, e \in \bigcup \mathcal{F}_2 \mid \exists y_1 \subseteq x_1 .\, (\, y_1,\ e\,) \in x \,\}$$

for $x \in \mathcal{F}_1 \to \mathcal{F}_2$, and

$$(\, y_1,\ e_2\,) \in \mu\, g \iff e_2 \in g(\, y_1\,) \ \&\ [y_1' \subseteq y_1 \ \&\ e_2 \in g(\, y_1'\,) \Longrightarrow y_1 = y_1']$$

for $g \in [\mathcal{F}_1 \to_s \mathcal{F}_2]$.

$Pt\, x$ is continuous because for any $(\, y,\ e\,) \in x$,y is finite. Let $x_1 \uparrow x_1'$ in $\mathcal{F}_1$. Then

$$\{\, e \in \bigcup \mathcal{F}_2 \mid \exists y_1 \subseteq x_1 .\, (\, y_1,\ e\,) \in x \,\}$$

and

$$\{\, e \in \bigcup \mathcal{F}_2 \mid \exists y_1 \subseteq x_1' .\, (\, y_1,\ e\,) \in x \,\}$$

are compatible in $\mathcal{F}_2$. It follows that

$$Pt\, x\, (x_1 \cap x_1') = Pt\, x\, (x_1) \cap Pt\, x\, (x_1').$$

Hence $Pt\, x$ is stable. One can show further that, for x, $y \in \mathcal{F}_1 \to \mathcal{F}_2$, $x \subseteq y$ if and only if $Pt\, x \sqsubseteq_s Pt\, y$.

For the other half of the pair, one can check that $\mu\, g$ is a member of $\mathcal{F}_1 \to \mathcal{F}_2$, and $g = Pt(\, \mu\, g\,)$.

□

6.2 Categories for Concurrency

This section presents the category $\mathbf{SEV}^*_{syn}$ and the category $\mathbf{SEV}_{syn}$. They were used to give models for concurrent languages like CCS. $\mathbf{SEV}^*_{syn}$ is a category with stable event structures as its objects and the *partially synchronous morphisms* as morphisms. $\mathbf{SEV}_{syn}$ is a category with stable event structures as its objects and the *synchronous morphisms* as morphisms.

Definition 6.16 *Let* $\underline{E}_1 = (E_1, Con_1, \vdash_1)$ *and* $\underline{E}_2 = (E_2, Con_2, \vdash_2)$ *be stable event structures. A partially synchronous morphism from* $\underline{E}_1$ *to* $\underline{E}_2$ *is a partial function* $\theta : E_1 \to E_2$ *on on events which satisfies*

- $X \in Con_1 \Longrightarrow \theta X \in Con_2$,
- $(\{e, e'\} \in Con_1 \ \&\ \theta(e) = \theta(e')) \Longrightarrow e = e'$, *and*
- $(X \vdash_1 e \ \&\ \theta(e) \text{ is defined }) \Longrightarrow \theta X \vdash_2 \theta(e)$.

A partially synchronous morphism θ *is synchronous if it is a total function.*

Note the truth of the equality $\theta(e) = \theta(e')$ asserts also that $\theta(e)$ and $\theta(e')$ are defined.

Definition 6.17 *Let* $\underline{E}_1 = (E_1, Con_1, \vdash_1)$ *and* $\underline{E}_2 = (E_2, Con_2, \vdash_2)$ *be stable event structures. Their partially synchronous product,* $\underline{E}_1 \times \underline{E}_2$, *is the structure* $(E, Con, \vdash)$ *where*

- $E = \{(e, *) \mid e \in E_1\} \cup \{(*, e') \mid e' \in E_2\}$
  $\cup\{(e, e') \mid e \in E_1 \ \&\ e' \in E_2\}$,
- $X \in Con \Longleftrightarrow \pi_1 X \in Con_1 \ \&\ \pi_2 X \in Con_2 \ \&$
  $\forall e, e' \in X. [\pi_1(e) = \pi_1(e') \text{ or } \pi_2(e) = \pi_2(e')] \Rightarrow e = e'$,
- $X \vdash e \Longleftrightarrow [\pi_1(e) \text{ is defined } \Rightarrow \pi_1 X \vdash_1 \pi_1(e)]\ \&$
  $[\pi_2(e) \text{ is defined } \Rightarrow \pi_2 X \vdash_2 \pi_2(e)]$.

An event $(e_1, *)$ is one which can occur independently of the events of $\underline{E}_2$. A pair of events (e_1, e_2) is understood as the synchronization of the event e_1 from $\underline{E}_1$ and e_2 from $\underline{E}_2$. The consistent predicate for the partially synchronous product indicates that in any computation, an event cannot synchronize with two distinct events.

Theorem 6.7 *The partially synchronous product with projections* π_1 *and* π_2 *is a product in the category* $\mathbf{SEV}^*_{syn}$.

Definition 6.18 *Let* $\underline{E}_1 = (E_1,\ Con_1,\ \vdash_1)$ *and* $\underline{E}_2 = (E_2,\ Con_2,\ \vdash_2)$ *be stable event structures. Their synchronous product,* $\underline{E}_1 \oplus \underline{E}_2$*, is the structure* $(E,\ Con, \vdash)$ *where*

- $E = \{(e,\ e') \mid e \in E_1\ \&\ e' \in E_2\}$,
- $X \in Con \iff \pi_1 X \in Con_1\ \&\ \pi_2 X \in Con_2\ \&$
  $\forall e,\ e' \in X.[\pi_1(e) = \pi_1(e')\ \text{ or }\ \pi_2(e) = \pi_2(e')] \Rightarrow e = e'$,
- $X \vdash e \iff (\pi_1 X \vdash_1 \pi_1(e)\ \&\ \pi_2 X \vdash_2 \pi_2(e))$.

Theorem 6.8 *The synchronous product with projections* π_1 *and* π_2 *is a product in the category* $\mathbf{SEV}_{syn}$.

Stable event structures can be used to give semantics to languages like CCS and CSP [Wi82], with the parallel composition modeled by the partially synchronous product and the nondeterministic choice modeled by the sum construction given below.

Definition 6.19 *Let* $\underline{E}_1 = (E_1,\ Con_1,\ \vdash_1)$ *and* $\underline{E}_2 = (E_2,\ Con_2,\ \vdash_2)$ *be stable event structures. Their sum,* $\underline{E}_1 + \underline{E}_2$*, is the structure* $(E,\ Con, \vdash)$ *where*

- $E = \{(1,\ e) \mid e \in E_1\} \cup \{(2,\ e) \mid e \in E_2\}$,
- $X \in Con \iff \exists X_1 \in Con_1.\ X = \{(1,\ e) \mid e \in X_1\}\ \text{ or }$
  $\exists X_2 \in Con_2.\ X = \{(2,\ e) \mid e \in X_2\}$,
- $X \vdash e \iff \exists X_1 \in Con_1,\ e_1 \in E_1.\ X_1 \vdash_1 e_1\ \&$
  $X = \{(1,\ e') \mid e' \in X_1\}\ \&\ e = (1,\ e_1)$
  or
  $\exists X_2 \in Con_2,\ e_2 \in E_2.\ X_2 \vdash_2 e_2\ \&$
  $X = \{(2,\ e') \mid e' \in X_2\}\ \&\ e = (2,\ e_2)$.

Theorem 6.9 *The sum is a coproduct in both the categories* $\mathbf{SEV}^*_{syn}$ *and* $\mathbf{SEV}_{syn}$.

6.3 Monoidal Closed Categories

There are two typical monoidal closed categories related to stable domains: $\mathbf{COH}_l$ and $\mathbf{SEV}_l$. The monoidal closed category of coherent spaces was used for a semantics of linear logic [Gi87b]. In fact, it is in this category that Girard discovered linear logic, based on a crucial observation that the stable function space can be decomposed into more elementary linear constructions.

6.3.1 $\mathbf{COH}_l$

Coherent families with linear, stable functions form a monoidal closed category $\mathbf{COH}_l$. The construction of tensor product and the linear function space are given by

Definition 6.20 *(Tensor Product) Let $\mathcal{F}_1$, $\mathcal{F}_2$ be coherent families. Their tensor product, $\mathcal{F}_1 \otimes \mathcal{F}_2$, is the family of subsets which satisfies*

$$x \in \mathcal{F}_1 \otimes \mathcal{F}_2 \Longleftrightarrow \quad \begin{array}{l} \bullet\ x \subseteq \bigcup \mathcal{F}_1 \times \bigcup \mathcal{F}_2, \\ \bullet\ \pi_1 x \in \mathcal{F}_1 \ \&\ \pi_2 x \in \mathcal{F}_2. \end{array}$$

Suppose $\mathcal{F}_1$ and $\mathcal{F}_2$ are coherent families and $y \subseteq x \in \mathcal{F}_1 \otimes \mathcal{F}_2$. Then clearly $y \subseteq \bigcup \mathcal{F}_1 \times \bigcup \mathcal{F}_2$ and $\pi_1 y \subseteq \pi_1 x$, $\pi_2 y \subseteq \pi_2 x$. Therefore $\pi_1 y \in \mathcal{F}_1$, $\pi_2 y \in \mathcal{F}_2$, and hence $y \in \mathcal{F}_1 \otimes \mathcal{F}_2$. This means $\mathcal{F}_1 \otimes \mathcal{F}_2$ is again a coherent family (the pairwise completeness is trivial).

Example 6.3 *Consider the product and tensor product of the following two coherent spaces.*

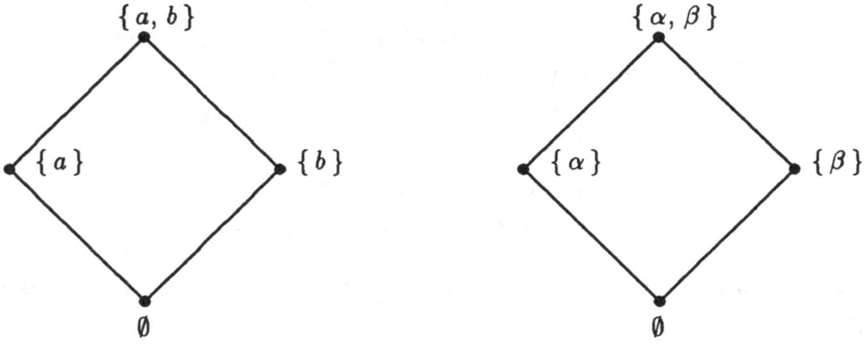

$\{(a, \alpha), (a, \beta), (b, \alpha)\}$ and $\{(a, \alpha), (a, \beta), (b, \beta)\}$ are different elements in the tensor product. Nevertheless, they have the same projections on the first component and the second component. Thus the tensor product has far more elements than the product.

Definition 6.21 *(Linear Function Space) Let $\mathcal{F}_1$, $\mathcal{F}_2$ be coherent families. Their linear function space, $\mathcal{F}_1 \multimap \mathcal{F}_2$, is the family of subsets which* *satisfies*

$$x \in \mathcal{F}_1 \multimap \mathcal{F}_2 \iff$$
$$\bullet\; x \subseteq \bigcup \mathcal{F}_1 \times \bigcup \mathcal{F}_2,$$
$$\bullet\; \forall y \subseteq x.\; \pi_1 y \in \mathcal{F}_1 \Longrightarrow \pi_2 y \in \mathcal{F}_2, \quad \textit{and}$$
$$\bullet\; [\{e_1, e_1'\} \in \mathcal{F}_1 \;\&\; (e_1, e_2),\; (e_1', e_2) \in x] \Longrightarrow e_1 = e_1'.$$

Any element x in $\mathcal{F}_1 \multimap \mathcal{F}_2$ determines a linear stable function $Pt\, x$, where

$$Pt\, x\,(x_1) = \{\, e_2 \in \bigcup \mathcal{F}_2 \mid \exists e_1 \in x_1.\,(e_1, e_2) \in x \,\}.$$

Linear function space is coherent, and is isomorphic to the linear functions from $\mathcal{F}_1$ to $\mathcal{F}_2$. The proof is very similar to the case of stable function space. Notice that for any $x \in \mathcal{F}_1 \multimap \mathcal{F}_2$, $i\,(x) \in \mathcal{F}_1 \to \mathcal{F}_2$, where

$$i\,(x) = \{\, (\{e_1\}, e_2) \mid (e_1, e_2) \in x \,\}.$$

COH$_l$ is a monoidal closed category, with the tensor product and linear function space introduced above. In particular, we have the monoidal closedness, which says that for any $\mathcal{F}_1$ $\mathcal{F}_2$, and $\mathcal{F}_3$,

$$[\mathcal{F}_1 \otimes \mathcal{F}_2 \multimap \mathcal{F}_3] \cong [\mathcal{F}_1 \multimap [\mathcal{F}_2 \multimap \mathcal{F}_3]].$$

6.3.2 SEV$_l$

If we change the definition of stable maps between stable event structures slightly by restricting the finite configuration x in an event (x, e) to be a complete prime $\lceil e \rceil_y$, we get the linear maps, the morphisms for the category **SEV**$_l$.

Definition 6.22 *Let* $\underline{E} = (E, Con_{\underline{E}}, \vdash_{\underline{E}})$, $\underline{F} = (F, Con_{\underline{F}}, \vdash_{\underline{F}})$ *be stable event structures. A linear map from* $\underline{E}$ *to* $\underline{F}$ *is a relation* $R \subseteq P_{\underline{E}} \times F$ *which satisfies*

1. *Compatibility* : $T \Subset R \;\&\; \bigcup(\pi_1 T) \in Con_{\underline{E}} \Longrightarrow \pi_2 T \in Con_{\underline{F}}$,
2. *Minimality* : $(x \cup y \in \mathcal{F}(\underline{E}) \;\&\; x\,R\,e \;\&\; y\,R\,e) \Longrightarrow x = y$, *and*
3. *Completeness:* $x\,R\,e \Longrightarrow \exists X.(X \vdash_{min} e \;\&\; \forall e' \in X \exists x' \subseteq x.\; x'\,R\,e')$.

Here $P_{\underline{E}}$ *stands for the set of complete primes of* $\mathcal{F}(\underline{E})$.

Stable event structures with linear maps form a category. There are the constructions of tensor product and linear function space which make $\mathbf{SEV}_l$ a monoidal closed category.

Definition 6.23 *Let $\underline{E}_1 = (E_1, Con_1, \vdash_1)$ and $\underline{E}_2 = (E_2, Con_2, \vdash_2)$ be stable event structures. Their tensor product, $\underline{E}_1 \otimes \underline{E}_2$, is defined as the structure $(E, Con, \vdash,)$ where*

$$E = E_1 \times E_2$$
$$X \in Con \Longleftrightarrow \pi_1 X \in Con_1 \ \&\ \pi_2 X \in Con_2$$
$$X \vdash e \Longleftrightarrow \exists X_1, X_2.\ X_1 \vdash_1 \pi_1 e \ \&\ X_2 \vdash_2 \pi_2 e \ \&\ X_1 \times X_2 \subseteq X.$$

Proposition 6.4 *If $\underline{E}_1$ and $\underline{E}_2$ are stable then so is $\underline{E}_1 \otimes \underline{E}_2$.*

Proof We check the stability axiom. Assume $X \vdash e$, $Y \vdash e$, and $X \cup Y \cup \{e\} \in Con$. By definition, there exist X_1, X_2, Y_1, and Y_2, such that

$$X_1 \vdash_1 \pi_1 e, \qquad X_2 \vdash_2 \pi_2 e,$$
$$Y_1 \vdash_1 \pi_1 e, \qquad Y_2 \vdash_2 \pi_2 e,$$
$$X \supseteq X_1 \times X_2, \quad Y \supseteq Y_1 \times Y_2.$$

Also we have $X_1 \cup Y_1 \cup \{\pi_1 e\} \in Con_1$ and $X_2 \cup Y_2 \cup \{\pi_2 e\} \in Con_2$. Therefore $X_1 \cap Y_1 \vdash_1 \pi_1 e$ and $X_2 \cap Y_2 \vdash_2 \pi_2 e$, by the stability of $\underline{E}_1$ and $\underline{E}_2$. However, $X \cap Y \supseteq (X_1 \cap Y_1) \times (X_2 \cap Y_2)$. Hence $X \cup Y \vdash e$.

□

Let $x \in \mathcal{F}(\underline{E}_1 \otimes \underline{E}_2)$. We have $\pi_1 x \in \mathcal{F}(\underline{E}_1)$. In fact, $\pi_1 x$ is clearly consistent and, if $e_1 \in \pi_1 x$, then there is some e_2 such that $(e_1, e_2) \in x$. The way enabling is defined for tensor product ensures that when (e_1, e_2) is secured in x, e_1 is secured in $\pi_1 x$. Hence $\pi_1 x$ is a configuration of $\underline{E}_1$. Similarly $\pi_2 x \in \mathcal{F}(\underline{E}_2)$. From this we know that for any (e_1, e_2) in x, $\lceil e_1 \rceil_{\pi_1 x} \times \lceil e_2 \rceil_{\pi_2 x} \subseteq x$.

Suppose, on the other hand, that x is a subset of $E_1 \times E_2$ with the property that $\pi_1 x \in \mathcal{F}(\underline{E}_1)$, $\pi_2 x \in \mathcal{F}(\underline{E}_2)$, and

$$\forall (e_1, e_2) \in x \exists x_1 \in \mathcal{F}(\underline{E}_1),\ x_2 \in \mathcal{F}(\underline{E}_2).\ (e_1, e_2) \in x_1 \times x_2 \subseteq x.$$

x is obviously a consistent set. Also, since $e_1 \in x_1$ and $e_2 \in x_2$,

$$\lceil e_1 \rceil_{\pi_1 x} \times \lceil e_2 \rceil_{\pi_2 x} \subseteq x_1 \times x_2 \subseteq x.$$

Hence (e_1, e_2) is secured in x. In summary, we have proved that

Proposition 6.5 *Let $\underline{E}_1 = (E_1, Con_1, \vdash_1)$ and $\underline{E}_2 = (E_2, Con_2, \vdash_2)$ be stable event structures. We have $x \in \mathcal{F}(\underline{E}_1 \otimes \underline{E}_2)$ if and only if*

- $\pi_1 x \in \mathcal{F}(\underline{E}_1)$ & $\pi_2 x \in \mathcal{F}(\underline{E}_2)$ *and*
- $\forall (e_1, e_2) \in x \exists x_1 \in \mathcal{F}(\underline{E}_1),\ x_2 \in \mathcal{F}(\underline{E}_2).\ (e_1, e_2) \in x_1 \times x_2 \subseteq x.$

Definition 6.24 *Let $\underline{E}_1 = (E_1, Con_1, \vdash_1)$ and $\underline{E}_2 = (E_2, Con_2, \vdash_2)$ be stable event structures. Their linear function space is defined as the structure $(E, Con, \vdash)$ where*

$$
\begin{aligned}
&E = \{(\lceil e_1 \rceil_x, e_2) \mid e_1 \in x \in \mathcal{F}(\underline{E}_1),\ e_2 \in E_2\},\\
&X \in Con \ \ \text{if} \forall Y \subseteq X.\ \bigcup(\pi_1 Y) \in Con_1 \Longrightarrow \pi_2 Y \in Con_2 \ \& \\
&\qquad\qquad \forall a, b \in X.(\pi_1 a \uparrow \pi_1 b \ \&\ \pi_2 a = \pi_2 b) \Longrightarrow a = b,\\
&X \vdash (x, e_2) \ \ \text{if}\ \{e \mid (y, e) \in X \ \&\ y \subseteq x\} \vdash_2 e_2.
\end{aligned}
$$

We write $\underline{E}_1 \multimap \underline{E}_2$ for the linear function space.

That the construction of linear function space is indeed what we wanted (given the linear maps) is confirmed in the following theorem.

Theorem 6.10 *The linear function space of two stable event structures is a stable event structure. R is a configuration in $\mathcal{F}[\underline{E}_1 \multimap \underline{E}_2]$ if and only if $R : \underline{E}_1 \to \underline{E}_2$ is a linear map.*

Proof The proof for the first part is routine.

To prove the second part, let R be a configuration in $\mathcal{F}[\underline{E}_1 \multimap \underline{E}_2]$. It is clear that R is compatible and minimal. Assume $(x, e) \in R$. Then (x, e) must be secured. By the definition of the enabling in the linear function space, this means, in particular, that e is secured in $\pi_2 R'$, the projection of a subset of R into the second component. Further more, the first components of R' must be contained in x. This shows R is complete and hence a linear map.

By the definition of the enabling relation of the linear function space, a linear map is a configuration.

□

One can show further that there is a 1-1, order preserving correspondence between the linear maps and linear (stable) functions. That is why the linear maps are called 'linear'.

Theorem 6.11 *There is a 1-1, order preserving correspondence between*

$$\{ f \mid f : \mathcal{F}(\underline{E}_1) \to \mathcal{F}(\underline{E}_2) \ \textit{ is linear, stable} \}$$

and

$$\{ R \mid R : \underline{E}_1 \to \underline{E}_2 \ \textit{ is a linear map} \}.$$

We also have

Proposition 6.6 *Let $\underline{E}_1$, $\underline{E}_2$ and $\underline{E}_3$ be stable event structures. Then*

$$(\underline{E}_1 \otimes \underline{E}_2) \otimes \underline{E}_3 \cong \underline{E}_1 \otimes (\underline{E}_2 \otimes \underline{E}_3).$$

This statement is easy to justify. It indicates that the tensor product is associative up to an isomorphism.

Theorem 6.12 *(Monoidal Closedness) Let $\underline{E}_1$, $\underline{E}_2$, and $\underline{E}_3$ be stable event structures. Then*

$$(\underline{E}_1 \otimes \underline{E}_2) \multimap \underline{E}_3 \cong \underline{E}_1 \multimap [\underline{E}_2 \multimap \underline{E}_3].$$

The isomorphism $\underline{E} \cong \underline{E}'$ amounts to saying that there is a 1-1 correspondence between E and E', which preserves the consistency and the entailment relations. A more categorical proof of monoidal closedness requires one to follow the definitions in [Ma71]. One can achieve such rigor but that is not the purpose here.

Proof The isomorphism is given by

$$\theta : \ (\lceil (e_1, e_2) \rceil_x, \ e_3) \longmapsto (\lceil e_1 \rceil_{\pi_1 x}, \ (\lceil e_2 \rceil_{\pi_2 x}, \ e_3)).$$

Clearly θ is onto as $x_1 \times x_2$ is a configuration in the tensor product if both x_1 and x_2 are. It is then clear that θ is a bijection. Write

$$\begin{aligned} (\underline{E}_1 \otimes \underline{E}_2) \multimap \underline{E}_3 &= (P, Con_P, \vdash_P), \\ \underline{E}_1 \multimap [\underline{E}_2 \multimap \underline{E}_3] &= (F, Con_F, \vdash_F), \ \text{ and} \\ [\underline{E}_2 \multimap \underline{E}_3] &= (E, Con_E, \vdash_E). \end{aligned}$$

Directly following the definitions one can check that $X \in Con_P$ if and only if $\theta X \in Con_F$, and $X \vdash_P e$ if and only if $\theta X \vdash_F \theta e$.

□

There is a similar monoidal closed category **DL** of dI-domains with linear, stable functions. It is a direct translation of the situation of the monoidal closed category $\mathbf{SEV}_l$ of stable event structures. It is easy to see that $\mathbf{SF}_l$, stable families with linear, stable functions also a monoidal closed category.

6.4 Relationships among the Categories

It is known that **DI**, $\mathbf{SEV}_s$, and $\mathbf{SF}_s$ are equivalent categories. It is also clear that **DL**, $\mathbf{SEV}_l$, and $\mathbf{SF}_l$ are equivalent categories. Apparently $\mathbf{SEV}_{syn}$ is a subcategory of $\mathbf{SEV}^*_{syn}$. Again, $\mathbf{SEV}_l$ is a subcategory of $\mathbf{SEV}_s$. $\mathbf{COH}_l$ is a full-subcategory of $\mathbf{SF}_l$; $\mathbf{COH}_s$ is a full-subcategory of $\mathbf{SF}_s$. There are other interesting relationships between some of the categories. In this section we show that there is an adjunction between $\mathbf{COH}_l$ and $\mathbf{COH}_s$.

Recall that one way to determine an adjunction between two categories **A** and **B**. Two functors $F : \mathbf{A} \to \mathbf{B}$, $G : \mathbf{B} \to \mathbf{A}$ is an adjunction pair if for any object a of **A**, there is a morphism $\Theta_a : a \to GF(a)$ in **A** which is *universal* in the sense explained as follows. $\Theta_a : a \to GF(a)$ in **A** is universal if for any morphism $f : a \to G(b)$ in **A** with b in **B** there is a unique morphism $h : F(a) \to b$ in **B** such that the following diagram commutes:

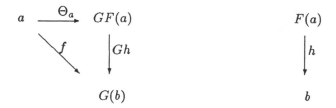

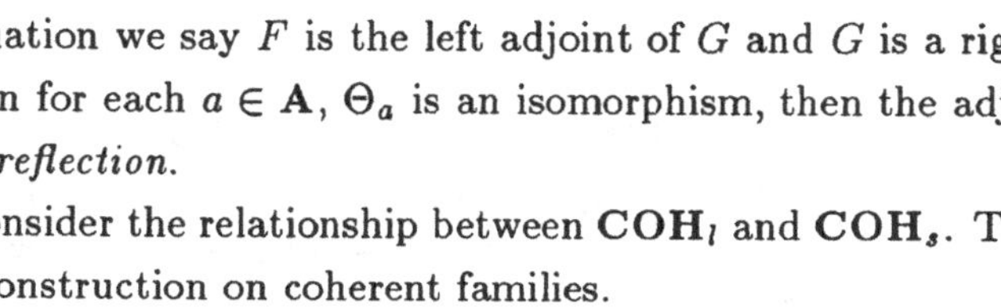

In this situation we say F is the left adjoint of G and G is a right adjoint of F. When for each $a \in \mathbf{A}$, Θ_a is an isomorphism, then the adjunction is called a *coreflection*.

First consider the relationship between $\mathbf{COH}_l$ and $\mathbf{COH}_s$. There is the following construction on coherent families.

Definition 6.25 *(Exponential) Let $\mathcal{F}$ be a coherent family. Its exponen-*

tial, $!\mathcal{F}$, *is the family of subsets which satisfies*

$$x \in !\mathcal{F} \iff x \subseteq \{ a \mid a \in \mathcal{F} \ \& \ | a | < \infty \} \ \& \ \bigcup x \in \mathcal{F}.$$

It is obvious that if $\mathcal{F}$ is a coherent family then $!\mathcal{F}$ is still a coherent family. The requirement that x consists of the *finite* configurations of $\mathcal{F}$ ensures that there are countably many events in $!\mathcal{F}$.

Clearly the inclusion i is a functor form $\mathbf{COH}_l$ to $\mathbf{COH}_s$. It has a left adjoint !, which sends a morphism $\mathcal{F}_1 \xrightarrow{g} \mathcal{F}_2$ in $\mathbf{COH}_s$ to a morphism $!\mathcal{F}_1 \xrightarrow{!g} !\mathcal{F}_2$ in $\mathbf{COH}_l$, where ! is the exponential operation on coherent families and $!g$ is a stable function determined by the linear map

$$\{(x, y) \mid | x | < \infty \ \& \ | y | < \infty \ \& \ y \subseteq g(x) \ \& \ \forall x' \subseteq x . y \subseteq g(x') \Rightarrow x = x'\}.$$

$\Theta_{\mathcal{F}} : \mathcal{F} \to !\mathcal{F}$ is given by $\Theta_{\mathcal{F}}(x) = \{ y \mid y \Subset x \}$. By inspecting the relevant axioms one can convince himself that ! is indeed a functor.

We check that $\Theta_{\mathcal{F}} : \mathcal{F} \to !\mathcal{F}$ is universal. For any morphism

$$f : \mathcal{F} \to \mathcal{F}'$$

in $\mathbf{COH}_s$, $h : !\mathcal{F} \to \mathcal{F}'$ is a morphism in $\mathbf{COH}_l$, where

$$h = Pt(\{ (x, e) \in \bigcup !\mathcal{F} \times \bigcup \mathcal{F}' \mid (x, e) \in \mu f \}).$$

Note that x has two different types in the definition of h. The first x is considered as an event of $!\mathcal{F}$, the second is considered as a configuration in $\mathcal{F}$. Clearly $h \circ \Theta_{\mathcal{F}} = f$, and such h is unique because it is linear. Thus the following diagram gives the universal property of $\Theta_{\mathcal{F}} : \mathcal{F} \to !\mathcal{F}$ where we omitted the inclusion functor i.

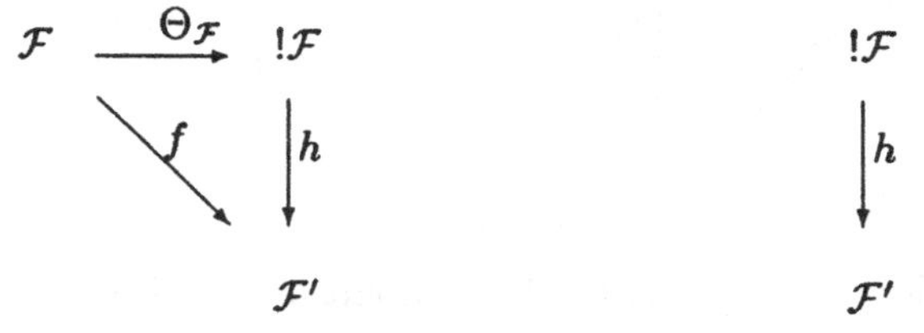

We have already known that any stable function $f : \mathcal{F} \to \mathcal{F}'$ gives a linear function $Pt(\{ (x, e) \in \bigcup !\mathcal{F} \times \bigcup \mathcal{F}' \mid (x, e) \in \mu g \})$ form $!\mathcal{F}$ to $\mathcal{F}'$. On the other hand, assume that $h : !\mathcal{F} \to \mathcal{F}'$ is a morphism in $\mathbf{COH}_l$. Then $\mu h \in !\mathcal{F} \multimap \mathcal{F}'$. Let

$$H = Pt(\{ (x, e) \mid (\{ x \}, e) \in \mu h \}).$$

It is easy to check that $\{(x, e) \mid (\{x\}, e) \in \mu h\}$ is a configuration in $\mathcal{F} \to \mathcal{F}'$, and

$$!\mathcal{F} \multimap \mathcal{F}' \cong \mathcal{F} \to \mathcal{F}'.$$

In summary, we have proved that

Theorem 6.13 $\mathbf{COH}_l \xrightarrow{i} \mathbf{COH}_s \xrightarrow{!} \mathbf{COH}_l$ *determines an adjunction, with !, inclusion, the left adjoint of i.*

There is also a construction of exponential in the categories $\mathbf{SEV}_l$ and **DL**. For a dI-domain D, its exponential is a dI-domain E such that there exists a 1-1, order preserving correspondence

$$! : D^0 \to E^p.$$

One way to construct such an exponential is to take the elements as the unions of finite downwards-closed subsets of D, and take the order as the inclusion. We therefore have

$$\forall a \in D^0.\ !a = \downarrow a.$$

This construction can be used to decompose stable functions to linear functions in exactly the same sense as what happened to coherent spaces. Surprisingly, however, the exponential we introduce here is not a generalization of the one on coherent spaces. Consider the exponential of the coherent space with a single token. Below, the picture on the right is the result of applying the exponential for coherent spaces, and the one on the left is the one on dI-domains–It is simply not a coherent space!

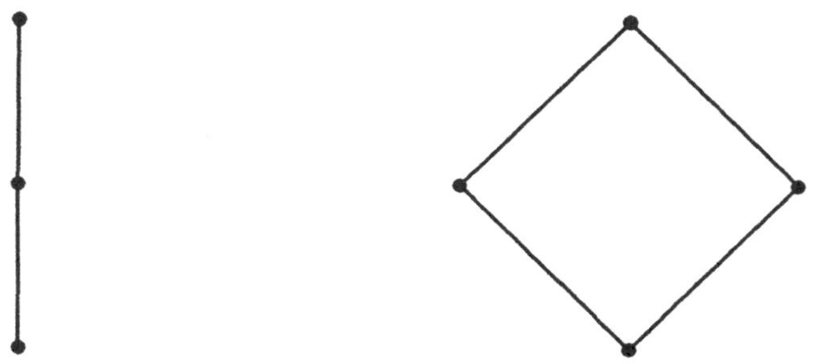

For more detail about the category $\mathbf{SEV}_l$ see [Zh91].

We summarize the known relationships among the categories by the following diagram:

$$\begin{array}{ccccccc}
\mathbf{COH}_s & \overset{1}{\longrightarrow} & \mathbf{SF}_s & \overset{2}{\longleftrightarrow} & \mathbf{DI} & \overset{2}{\longleftrightarrow} & \mathbf{SEV}_s \\
\updownarrow 3 & & \updownarrow 3 & & \updownarrow 3 & & \updownarrow 3 \\
\mathbf{COH}_l & \overset{1}{\longrightarrow} & \mathbf{SF}_l & \overset{2}{\longleftrightarrow} & \mathbf{DL} & \overset{2}{\longleftrightarrow} & \mathbf{SEV}_l \\
 & & & & & & \uparrow 4 \\
 & & & & & & \mathbf{SEV}^*_{syn} \\
 & & & & & & \uparrow 4 \\
 & & & & & & \mathbf{SEV}_{syn}
\end{array}$$

where the labels are interpreted as:

1 : full-subcategory,
2 : equivalent,
3 : adjunction,
4 : subcategory.

However, that shouldn't be the end of the story. Other relations among the categories may exist. It is believed that, for example, there is a coreflection between $\mathbf{SEV}_l$ and $\mathbf{COH}_l$ with the double linear negation as one of the functors.

Chapter 7

A Representation of DI

We know from the previous chapter that dI-domains can be represented as stable event structures. It may have been observed by the reader that there is some similarity between event structures and information systems. The difference is, however, perhaps bigger than the similarity after a second thought. For an information system, for example, if $X \vdash a$ and $X \vdash a'$ then a and a' must be consistent while for an event structure, we cannot say anything about the consistency of two events e, e' enabled by the same set of events. This reflects the fact that for information systems $\vdash$ stands for logical entailment between propositions whereas in the case of event structures it expresses sufficient conditions for an event to occur due to the previous occurrences of other events. Thus information systems capture the logical relations between facts about a computation while event structures capture their temporal relationship.

This chapter introduces prime information systems, to represent dI-domains by logical structures similar to information systems. It provides an alternative account of dI-domains, as well as a foundation for recursively defined systems. With respect to the topic of Part II, it's role is to establish the backbones of the logic of dI-domains. Therefore, this chapter plays a similar role as Chapter 3 did for Part I.

Section 7.1 gives the definition of prime information systems and shows that they determine dI-domains. Section 7.2 introduces a category of prime information systems. This category is shown equivalent to the category of dI-domains. In Section 7.3, a complete partial order of prime information system is given so that recursive prime information systems can be defined. Constructions in the category of prime information systems are introduced in Section 7.4. Various common constructions turn out specifying continuous functions on the complete partial order given in Section 7.3. Finally, a treatment of coherent spaces along this line is provided in Section 7.5.

7.1 Prime Information Systems

This section introduces prime information systems and shows that they determine dI-domains.

Definition 7.1 *An information system (see Definition 3.3)*

$$\underline{A} = (A, Con, \vdash)$$

is called prime if it satisfies two extra axioms:

6. $X \vdash a \implies \exists b \in X. \{ b \} \vdash a$,
7. $\{ b \mid X \vdash b \}$ *is finite for all finite proposition set* X.

Axiom 6 indicates that the entailment $\vdash$ is determined by a preorder on A by letting $a \le b$ if $\{ b \} \vdash a$. Thus prime information systems are similar to prime event structures. The only difference is that the entailment here determines a preorder while for prime event structures the causal dependency is a partial order. Thus prime event structures can be seen both as information systems and event structures. By taking a logical approach (tokens as propositions and configurations as theories) we committed ourselves to not requiring that the entailment on propositions be a partial order. The technical differences between prime information systems and prime event structures are superficial at this stage, though. However, later when it comes to the constructions in the category of prime information systems, especially the construction of stable function space, the reason for taking the logical stand (and hence the name prime information systems instead of prime event structures) becomes more clear. There, molecules are used as tokens. They are finite collections of pairs which specify some minimal information about entailment. It turns out necessary to have a collection of pairs, regarded as a conjunction of specifications, to determine the minimal information. Therefore regarding the molecules as events seemed to lack a computational justification.

Axiom 7 corresponds to the axiom of finite causes for event structures. There is a strong computational intuition behind that axiom in [Wi88]. Another justification for Axiom 7 is to get a cartesian closed category with A countable, as pointed out earlier in the discussion after Lemma 6.4. Note that a choice is made here: we could have used an axiom which requires that there be only finitely many $\approx$-equivalent classes in $\{ b \mid X \vdash b \}$ rather than the whole set $\{ b \mid X \vdash b \}$ be finite.

Axiom 7 is equivalent to, under the assumption of Axiom 6, the following

$$\text{Axiom 7'} : \forall a \in A.\ \{ b \mid \{ a \} \vdash b \} \text{ is finite,}$$

which is easier to use. Obviously Axiom 7 implies Axiom 7'. In reverse, assume we have Axiom 7'. By Axiom 6, whenever $X \vdash b$ there is some $a \in X$ such that $\{ a \} \vdash b$. However, for each $a \in X$, $\{ b \mid \{ a \} \vdash b \}$ is finite. Therefore $\{ b \mid X \vdash b \}$ is also finite, since X is.

Theorem 7.1 *Let* $\underline{A} = (A, Con, \vdash)$ *be a prime information system. Then* $(| \underline{A} |, \subseteq)$ *is a dI-domain.*

Proof Axiom 7 implies that $(| \underline{A} |, \subseteq)$ is finitary. By Theorem 6.2, for Axiom d to hold it is enough to show that elements of the form

$$\{ b \in A \mid \{ a \} \vdash_{\underline{A}} b \}$$

with $a \in A$ are the complete primes of $(| \underline{A} |, \subseteq)$, and $(| \underline{A} |, \subseteq)$ is prime algebraic. These, however, follow from Axiom 6.

□

On the other hand, given a dI-domain D, it is easy to construct a prime information system so that the domain determined by the system is isomorphic to D. All one has to do it to take the complete primes of D as propositions. More formally,

Definition 7.2 *Let D be a dI-domain. Define*

$$PI(D) = (A, Con, \vdash),$$

by taking

- $A = \{\uparrow p \mid p \in D^p\}$,
- $X \in Con \Longleftrightarrow \bigcap X \neq \emptyset$,
- $X \vdash a \Longleftrightarrow \bigcap X \subseteq a$.

Following standard convention, we let $\bigcap X = D$ when X is empty.

Proposition 7.1 *If D is a dI-domain then $PI(D)$ is a prime information system. Furthermore, there is a 1-1, order preserving correspondence between D and $PI(D)$.*

The proof is straightforward, hence omitted.

7.2 A Category of PIS

Note that to capture the stable order, a slightly modified version of the approximable mappings used in the category of information systems does not work. However, Theorem 6.3 suggests that we use

Definition 7.3 *Let $\underline{A} = (A, Con_{\underline{A}}, \vdash_{\underline{A}})$, $\underline{B} = (B, Con_{\underline{B}}, \vdash_{\underline{B}})$ be prime information systems. A stable approximable mapping $R : \underline{A} \to \underline{B}$ is a relation $R \subseteq |\underline{A}|^0 \times B$ which satisfies*

1. *Compatibility* : $(T \Subset R \ \& \ \bigcup \pi_1 T \in Con_{\underline{A}}) \Longrightarrow \pi_2 T \in Con_{\underline{B}}$,
2. *Minimality* : $(a \cup b \in |\underline{A}| \ \& \ a\,R\,p \ \& \ b\,R\,p) \Longrightarrow a = b$, *and*
3. *Completeness* : $(a\,R\,p \ \& \ \{p\} \vdash_{\underline{B}} q) \Longrightarrow b\,R\,q$ *for some* $b \subseteq a$.

In this definition the compatibility requirement is standard. The second condition expresses the *minimal* property. $a\,R\,p$ can be read as: a in $\underline{A}$ entails p in $\underline{B}$, and, moreover, a is a weakest one in $\underline{A}$ (or, it is not only sufficient but also necessary for a to produce p). For example, for any stable approximable mapping R, we cannot have $\emptyset\,R\,p$ and $a\,R\,p$ for some non-empty a at the same time. The third condition insists on *completeness*, in the sense that when a is a weakest proposition for p, all the propositions weaker than p must also have their weakest propositions specified.

Suppose $a\,R\,p$ and $\{p\} \approx \{q\}$. By the completeness axiom there is some $b \subseteq a$ such that $b\,R\,q$. But again, $\{q\} \vdash p$. Hence $c\,R\,p$ for some

$c \subseteq b$. Clearly $a \cup c$ is a finite element. By minimality we must have $a = c$. Therefore $a = b$, which means $a\, R\, q$, too. This analysis shows that under the completeness axiom, minimality is equivalent to the otherwise weaker statement

$$(a \cup b \in \mid \underline{A} \mid \ \&\ a\, R\, p\ \&\ b\, R\, q\ \&\ p \approx_{\underline{B}} q) \Longrightarrow a = b.$$

Proposition 7.2 *Suppose $R : \underline{A} \to \underline{B}$ is a stable approximable mapping. Then the function $Pt\,(R) : \mid \underline{A} \mid \to \mid \underline{B} \mid$ specified by*

$$Pt\,(R)(x) = \{\, p \mid \exists a \subseteq x \,.\, a\, R\, p \,\}$$

is stable.

Proof First we check $Pt\,(R)$ is well-defined.

For any element $x \in \mid \underline{A} \mid$, $Pt\,(R)(x)$ is finitely consistent. Suppose

$$\{\, p_0, p_1, \cdots, p_n \,\} \subseteq Pt\,(R)(x).$$

There exist $a_0, a_1, \cdots, a_n$, such that

$$\forall i\,.\,(\, a_i \subseteq x\ \&\ a_i\, R\, p_i \,).$$

By the compatibility axiom we have $\{\, p_0, p_1, \cdots, p_n \,\} \in Con_{\underline{B}}$.

$Pt\,(R)(x)$ is also closed under entailment. Assume $Y \subseteq Pt\,(R)(x)$ and $Y \vdash q$. We know that $\exists p \in Y\,.\,\{\, p \,\} \vdash q$. Since $p \in Pt\,(R)(x)$, there exists an $a \subseteq x$ such that $a\, R\, p$. By the completeness axiom $b\, R\, q$ for some $b \subseteq a$, which implies $q \in Pt\,(R)(x)$. Therefore $Pt\,(R)(x)$ is an element in $\mid \underline{B} \mid$.

It is routine to check that $Pt\,(R)$ is continuous. To check stability assume $x \uparrow y$ with $x, y \in \mid \underline{A} \mid$. Assume also that $p \in Pt\,R(x) \cap Pt\,R(y)$. Thus $\exists a \subseteq x\,, b \subseteq y$ such that $a\, R\, p$ and $b\, R\, p$. Since x and y are compatible, $a \cup b \in \mid \underline{A} \mid$. Therefore, by the minimality axiom, $a = b$, which implies $p \in Pt\,R(x \cap y)$. We have proved that $Pt\,R(x) \cap Pt\,R(y) \subseteq Pt\,R(x \cap y)$. The other direction of the inclusion follows from the monotonicity of $Pt\,R$.

□

Moreover, set inclusion on stable approximable mappings determines the stable order.

Proposition 7.3 *Let $\underline{A}$ and $\underline{B}$ be prime information systems, and*

$$R,\ S : \underline{A} \to \underline{B}$$

stable approximable mappings. Then $R \subseteq S$ *if and only if*

$$Pt(R) \sqsubseteq_s Pt(S),$$

where $\sqsubseteq_s$ *is the stable order.*

Proof Only if: Assume $R \subseteq S$ and let $x,\ y \in |\underline{A}|$, $x \subseteq y$ and

$$p \in Pt(R)(y) \cap Pt(S)(x).$$

There must be $a \subseteq x$ and $b \subseteq y$ such that $a\,R\,p$ and $b\,S\,p$. By a similar argument used in the proof of Proposition 7.2 we know that

$$Pt\,R(x) \supseteq Pt\,R(y) \cap Pt\,S(x).$$

The other direction of the inclusion follows from monotonicity.

If: Suppose $Pt(R) \sqsubseteq_s Pt(S)$. If $a\,R\,p$ then $p \in (Pt\,R)(a) \subseteq (Pt\,S)(a)$. Therefore $b\,S\,p$ for some $b \subseteq a$. By the stable order we get

$$Pt\,R(b) = Pt\,R(a) \cap Pt\,S(b).$$

It is easy to deduce $p \in Pt\,R(a) \cap Pt\,S(b)$ since $a\,R\,p$ and $b\,S\,p$. This implies $p \in Pt\,R(b)$. For some $c \subseteq b$, therefore, $c\,R\,p$. By the compatibility axiom we have $a = c$, and therefore $a\,S\,p$.

□

Clearly the identity stable approximable mappings *Id* should be given by $a\,Id\,p$ if and only if $a \approx \{p\}$. However, to compose stable approximable mappings we have to introduce $\tilde{R}$ for a stable approximable mapping $R : \underline{A} \to \underline{B}$. Define $\tilde{R}$ to be a relation on $|\underline{A}|^0 \times |\underline{B}|^0$ such that $a\,\tilde{R}\,b$ iff there exist $a_i,\ p_i$ such that $a_i\,R\,p_i$, $1 \le i \le n$, $a = \bigcup_{1 \le i \le n} a_i \in Con_{\underline{A}}$, and $b = \{p_i \mid 1 \le i \le n\}$. It is not difficult to check that $\tilde{R}$ has similar properties to that of stable approximable mappings:

1. $H \Subset \tilde{R}\ \&\ \bigcup(\pi_1 H) \in Con_{\underline{A}} \Longrightarrow \pi_2 H \in Con_{\underline{B}}$,
2. $(a \cup b \in |\underline{A}|\ \&\ a\,\tilde{R}\,c\ \&\ b\,\tilde{R}\,c) \Longrightarrow a = b$, and
3. $a\,\tilde{R}\,b \supseteq c \Longrightarrow \exists a' \subseteq a.\, a'\,\tilde{R}\,c$.

Intuitively, $\tilde{R}$ extends the minimal relation from propositions in $\underline{B}$ to its finite elements. The composition of $R : \underline{A} \to \underline{B}$ and $S : \underline{B} \to \underline{C}$ can now be defined as $R \circ S$, which is a relation on $Con_{\underline{A}} \times C$ such that

$$a\,(R \circ S)\,p \Longleftrightarrow \exists b \in |\underline{A}|^0.\ a\,\tilde{R}\,b\ \&\ b\,S\,p.$$

From the property of $\tilde{R}$ one can see that $R \circ S : \underline{A} \to \underline{C}$ is a stable approximable mapping and further, by inspecting the axioms for a category, we know that prime information systems with stable approximable mappings form a category, written as **PIS**.

We can get stable approximable mappings from stable functions.

Definition 7.4 *Let D, E be dI-domains, and $f : D \to E$ a stable function. Define a relation $PI(f) \subseteq | PI(D) |^0 \times | PI(E) |^p$ by taking*

$$X\ PI(f)\ a \Longleftrightarrow (x, p) \in \mu f$$

provided $\bigcap X = \uparrow x$ and $a = \uparrow p$.

Note $(x, p) \in \mu f$ implies that x is a finite element of D and p is a complete prime of E. The proposition below, which is routine to prove, confirms what is expected.

Proposition 7.4 *Let D, E be dI-domains, and $f : D \to E$ a stable function. Then $PI(f)$ is a stable approximable mapping from $PI(D)$ to $PI(E)$.*

We conclude this section by

Theorem 7.2 *Pt : **PIS** $\to$ **DI** and PI : **DI** $\to$ **PIS** are functors which determine an equivalence of **PIS** and **DI**.*

Proof Here Pt is a functor which sends a prime information system $\underline{A}$ to the dI-domain $| \underline{A} |$ and a stable approximable mapping R to $Pt(R)$ as given in Proposition 7.2.

We use one of MacLane's results in [Ma71]. It is enough to show that Pt is full and faithful, and each dI-domain D is isomorphic to $| \underline{A} |$ for some prime information system $\underline{A}$. The latter is straightforward. It remains to show that Pt is full and faithful.

First we show that Pt is full. Suppose $\underline{A}$ and $\underline{B}$ are prime information systems and

$$f : | \underline{A} | \to | \underline{B} |$$

a stable function. Define a relation $R \subseteq | \underline{A} |^0 \times B$ by letting $a R p$ if

$$(a, \{ q \mid \{p\} \vdash_{\underline{B}} q \}) \in \mu f.$$

It follows from Proposition 7.4 that this relation is an approximable mapping form $\underline{A}$ to $\underline{B}$. By Theorem 6.3, the stable function $Pt\, R$ determined by R is actually equal to f.

Suppose $R,\ S : \underline{A} \to \underline{B}$ are approximable mappings such that $Pt\, R = Pt\, S$. It follows from Proposition 7.3 that $R = S$. Therefore Pt is faithful.

□

7.3 A Cpo of PIS

This section introduces a subsystem relation on prime information systems. The subsystem relation captures the notion of a rigid embedding [KP78]. Under the subsystem relation prime information systems form a cpo, which makes it possible to give recursively defined prime information systems through the construction of least fixed points for continuous functions.

Definition 7.5 *Let $\underline{A} = (A, Con_{\underline{A}}, \vdash_{\underline{A}})$ and $\underline{B} = (B, Con_{\underline{B}}, \vdash_{\underline{B}})$ be prime information systems. $\underline{A} \trianglelefteq \underline{B}$ if*

1. $A \subseteq B$,
2. $X \in Con_{\underline{A}} \iff X \subseteq A \,\&\, X \in Con_{\underline{B}}$, *and*
3. $X \vdash_{\underline{A}} a \iff X \subseteq A \,\&\, X \vdash_{\underline{B}} a$.

When $\underline{A} \trianglelefteq \underline{B}$ we call $\underline{A}$ a ***rigid subsystem*** of $\underline{B}$. Note the third condition above implies that

$$X \vdash_{\underline{B}} a \,\&\, X \subseteq A \Longrightarrow a \in A.$$

Hence we have a stronger notion of subsystems than used in Definition 3.19, Section 3.6 (This is not surprising at all because rigid embeddings are embeddings but not vice versa).

Let $\underline{A} = (A, Con_{\underline{A}}, \vdash_{\underline{A}})$ and $\underline{B} = (B, Con_{\underline{B}}, \vdash_{\underline{B}})$ be prime information systems. If $A = B$ and $\underline{A} \trianglelefteq \underline{B}$, then $\underline{A} = \underline{B}$.

Definition 7.6 *Let D, E be dI-domains. A stable function $f : D \to E$ is a rigid embedding if there is a stable function $g : E \to D$ called a projection such that*

- $\forall d \in D.\ gf(d) = d$,
- $\forall e \in E.\ fg(e) \sqsubseteq e$, *and*
- $\forall d \in D,\ e \in E\,.\ e \sqsubseteq f(d) \Longrightarrow fg(e) = e$.

Proposition 7.5 *Let* $\underline{A} = (A, Con_{\underline{A}}, \vdash_{\underline{A}})$ *and* $\underline{B}$ *be prime information systems. If* $\underline{A} \trianglelefteq \underline{B}$ *then the inclusion map* $i : |\underline{A}| \rightarrow |\underline{B}|$ *is a rigid embedding with the projection* $j : |\underline{B}| \rightarrow |\underline{A}|$ *given by* $j(y) = y \cap A$ *for* $y \in |\underline{B}|$.

Proof We have

$$\forall x \in |\underline{A}| \,.\; ji(x) = x \cap A = x,$$
$$\forall y \in |\underline{B}| \,.\; ij(y) = y \cap A \subseteq y, \text{ and}$$
$$\forall y \in |\underline{B}| \,.\; y \subseteq A \Rightarrow ij(y) = y.$$

Hence it is enough to show that i, j are well-defined functions, which is trivial.

□

The relation $\trianglelefteq$ is almost a complete partial order on prime information systems. Clearly there is a least prime information system, which has an empty proposition set. The limit of an ω-increasing chain is a prime information system with the proposition set, consistency and entailment relations the union of those in the chain. We have

Theorem 7.3 *The relation* $\trianglelefteq$ *is a complete partial order with the least element*

$$\perp = (\emptyset, \{\emptyset\}, \emptyset).$$

If $\underline{A}_0 \trianglelefteq \underline{A}_1 \trianglelefteq \cdots \trianglelefteq \underline{A}_i \trianglelefteq \cdots$ *is an increasing chain of prime information systems where* $\underline{A}_i = (A_i, Con_i, \vdash_i)$, *then their least upper bound is*

$$\bigsqcup_{i\in\omega} \underline{A}_i = \left(\bigcup_{i\in\omega} A_i, \bigcup_{i\in\omega} \vdash_i, \bigcup_{i\in\omega} Con_i \right).$$

Proof As the notion of rigid subsystem is stronger than that of the subsystem we know that

$$\bigsqcup_{i\in\omega} \underline{A}_i = \left(\bigcup_{i\in\omega} A_i, \bigcup_{i\in\omega} \vdash_i, \bigcup_{i\in\omega} Con_i \right)$$

is the least upper bound, in the cpo of information systems. We check that Axiom 6 and Axiom 7 of Definition 7.1 hold for $\bigcup_{i\in\omega} \underline{A}_i$. Suppose $X \vdash a$ in $\bigcup_{i\in\omega} \underline{A}_i$. Since the entailment is the union of those of $\underline{A}_i$'s and X is finite, there must be some k such that $X \vdash_k a$. But $\underline{A}_k$ is prime. Therefore

$\exists b \in X. \{ b \} \vdash_k a$ which implies $\{ b \} \vdash a$ in $\bigcup_{i \in \omega} \underline{A}_i$. To see Axiom 7 holds consider $\{ b \mid \{ a \} \vdash b \}$ in $\bigcup_{i \in \omega} \underline{A}_i$. Obviously $a \in A_k$ for some k. We show that

$$\{ b \mid \{ a \} \vdash b \} \subseteq \{ b \mid \{ a \} \vdash_k b \},$$

which implies the finiteness of $\{ b \mid \{ a \} \vdash b \}$. Assume $\{ a \} \vdash_j t$ for $t \in A_j$. If $j \leq k$ then $\{ a \} \vdash_k t$ as $\underline{A}_j \trianglelefteq \underline{A}_k$. If $j \geq k$ then $\underline{A}_k \trianglelefteq \underline{A}_j$. By Axiom 3 of Definition 7.5 $\{ a \} \vdash_k t$ since $a \in A_k$. Therefore Axiom 7 holds for $\bigcup_{i \in \omega} \underline{A}_i$.

That for each j $\underline{A}_j \trianglelefteq \bigcup_{i \in \omega} \underline{A}_i$ is trivial.

□

Write $\mathbf{CPO}_{\text{pis}}$ for the 'cpo' of prime information systems under $\trianglelefteq$. $\mathbf{CPO}_{\text{pis}}$ is not a cpo in the usual sense simply because they are not a set but a class. However, this large cpo still suits our purpose.

The rigid-subsystem relation $\trianglelefteq$ can be easily extended to n-tuples coordinatewise. More precisely we require

$$(\underline{A}_1, \underline{A}_2, \cdots \underline{A}_n) \trianglelefteq (\underline{B}_1, \underline{B}_2, \cdots \underline{B}_n)$$

if for each $1 \leq i \leq n$, $\underline{A}_i \trianglelefteq \underline{B}_i$. For convenience write $\vec{\underline{A}}$ for $(\underline{A}_1, \underline{A}_2, \cdots \underline{A}_n)$.

The least upper bound of an ω-chain of n-tuples of prime information systems is then just the n-tuple of prime information systems consisting of the least upper bounds on each component, *i.e.* if

$$\vec{\underline{A}}_1 \trianglelefteq \vec{\underline{A}}_2 \cdots \trianglelefteq \vec{\underline{A}}_i \trianglelefteq \cdots$$

then for the j-th component

$$\pi_j \left(\bigsqcup_{i \in \omega} \vec{\underline{A}}_i \right) = \bigsqcup_{i \in \omega} \pi_j \left(\vec{\underline{A}}_i \right).$$

An operation F from n-tuples of prime information systems to m-tuples of prime information systems is said to be continuous if it is monotonic, *i.e.* $\vec{\underline{A}} \trianglelefteq \vec{\underline{B}}$ implies $F(\vec{\underline{A}}) \trianglelefteq F(\vec{\underline{B}})$, and preserves ω-increasing chains of prime information systems, *i.e.*

$$\vec{\underline{A}}_1 \trianglelefteq \vec{\underline{A}}_2 \cdots \trianglelefteq \vec{\underline{A}}_i \trianglelefteq \cdots$$

implies

$$\bigsqcup_{i \in \omega} F(\vec{\underline{A}}_i) = F(\bigsqcup_{i \in \omega} \vec{\underline{A}}_i).$$

It is well-known that for functions on (finite) tuples of cpos they are continuous if and only if by changing (any) one argument while fixing others the induced function is continuous.

Proposition 7.6 *An unary operation F is continuous if and only if it is monotonic with respect to* $\trianglelefteq$ *and continuous on proposition sets, i.e. for any ω-increasing chain*

$$\underline{A}_1 \trianglelefteq \underline{A}_2 \cdots \trianglelefteq \underline{A}_i \trianglelefteq \cdots,$$

each proposition of $F(\bigsqcup_{i\in\omega} \underline{A}_i)$ *is a proposition of* $\bigsqcup_{i\in\omega} F(\underline{A}_i)$.

Proof The 'only if' part is trivial.
If: Let

$$\underline{A}_1 \trianglelefteq \underline{A}_2 \cdots \trianglelefteq \underline{A}_i \trianglelefteq \cdots$$

be an ω- increasing chain of prime information systems. Since F is monotonic, we clearly have

$$\bigsqcup_{i\in\omega} F(\underline{A}_i) \trianglelefteq F(\bigsqcup_{i\in\omega} \underline{A}_i).$$

Thus the propositions of $F(\bigsqcup_{i\in\omega} \underline{A}_i)$ are the same as proposition of

$$\bigsqcup_{i\in\omega} F(\underline{A}_i).$$

Therefore they are the same prime information systems by the remark given just before Definition 7.6.

□

Now for any continuous function F on $\mathbf{CPO}_{\text{pis}}$, we can get the least fixed-point of F as the limit of the increasing ω-chain

$$\underline{\bot} \trianglelefteq F(\underline{\bot}) \trianglelefteq F^2(\underline{\bot}) \trianglelefteq \cdots \trianglelefteq F^n(\underline{\bot}) \trianglelefteq \cdots.$$

Note since we are working with a partial order, we get an equality

$$F(\bigsqcup_{i\in\omega} F^i(\underline{\bot})) = \bigsqcup_{i\in\omega} F^i(\underline{\bot}).$$

7.4 Constructions

In the previous section we have introduced a method to solve equations of prime information systems with the help of fixed-point theory. A necessary step before applying this method is to check that we indeed have some useful continuous functions. Without introducing the constructions, and hence a class of continuous functions, what we did in the previous section would be like shooting with no target. Therefore, we spend this section to discuss constructions. They include lifting $(\quad)_\uparrow$, sum $+$, product $\times$ and function space $\to$. These constructions have their counterparts in dI-domains as the constructions of lifting, sum, product and stable function space. By showing that they induce continuous functions on $\mathbf{CPO}_{\mathrm{pis}}$, we have made it possible to produce solutions to recursive equations in these constructions.

Lifting, sum and product are more or less the same as those given in Chapter 3.

Definition 7.7 *(Lifting) Let $\underline{A} = (A,\ Con,\ \vdash)$ be a prime information system. Define the lift of $\underline{A}$ to be $\underline{A}_\uparrow = (A',\ Con',\ \vdash')$ where*

- $A' = (\{0\} \times A) \cup \{0\}$,
- $X \in Con' \Longleftrightarrow \{a \mid (0,\ a) \in X\} \in Con$,
- $X \vdash' a \Longleftrightarrow [X \neq \emptyset \& a = 0 \text{ or } a = (0,\ b) \& \{c \mid (0,\ c) \in X\} \vdash b]$.

Lifting is an operation which given a prime information system produces a new one by joining a new proposition weaker than all the old ones.

Definition 7.8 *(Sum) Let*

$$\underline{A} = (A,\ Con_{\underline{A}},\ \vdash_{\underline{A}})$$

and

$$\underline{B} = (B,\ Con_{\underline{B}},\ \vdash_{\underline{B}})$$

be prime information systems. Define their sum,

$$\underline{A} + \underline{B},$$

to be $\underline{C} = (C, Con, \vdash)$ *where*

- $C = \{0\} \times A \cup \{1\} \times B$,
- $W \in Con \iff \exists X \in Con_{\underline{A}}.\ W = \{(0, a) \mid a \in X\}$ *or* $\exists Y \in Con_{\underline{B}}.\ W = \{(1, b) \mid b \in Y\}$,
- $W \vdash c \iff W = \{(0, a) \mid a \in X\}\ \&\ c = (0, r)\ \&\ X \vdash_{\underline{A}} r$ *or* $W = \{(1, b) \mid b \in Y\}\ \&\ c = (1, t)\ \&\ Y \vdash_{\underline{B}} t$.

The effect of sum is to juxtaposing disjoint copies of two prime information systems. We can obtain the separated sum $\oplus$ by letting $\underline{A} \oplus \underline{B} =^{\text{def}} \underline{A}_{\uparrow} + \underline{B}_{\uparrow}$.

Definition 7.9 *(Product) Let*

$$\underline{A} = (A, Con_{\underline{A}}, \vdash_{\underline{A}})$$

and

$$\underline{B} = (B, Con_{\underline{B}}, \vdash_{\underline{B}})$$

be prime information systems. Define their product, $\underline{A} \times \underline{B}$, *to be*

$$\underline{C} = (C, Con, \vdash)$$

where

- $C = \{0\} \times A \cup \{1\} \times B$,
- $W \in Con \iff \{a \mid (0, a) \in W\} \in Con_{\underline{A}}\ \&\ \{b \mid (1, b) \in W\} \in Con_{\underline{B}}$,
- $W \vdash c \iff c = (0, r)\ \&\ \{a \mid (0, a) \in W\} \vdash_{\underline{A}} r$ *or* $c = (1, t)\ \&\ \{b \mid (1, b) \in W\} \vdash_{\underline{B}} t$.

The proposition set of the product is the disjoint union of propositions of the components. A finite set of propositions is consistent if the projections to the components are. And a consistent set entails a proposition if it does so when projected into the appropriate component.

Definition 7.10 *Let* $\underline{A} = (A, Con_{\underline{A}}, \vdash_{\underline{A}})$ *and* $\underline{B} = (B, Con_{\underline{B}}, \vdash_{\underline{B}})$ *be prime information systems. A molecule* m *is a finite stable approximable mapping such that for some* $(a, p) \in m$, $b \subseteq a$ *and* $\{p\} \vdash q$ *for any other* (b, q) *in* m.

By Theorem 6.3, molecules capture complete primes in the function space.

Definition 7.11 *(Function Space) Let*

$$\underline{A} = (A, Con_{\underline{A}}, \vdash_{\underline{A}})$$

and

$$\underline{B} = (B, Con_{\underline{B}}, \vdash_{\underline{B}})$$

be prime information systems. Define their function space, $[\underline{A} \to \underline{B}]$, to be $\underline{C} = (C, Con, \vdash)$ where

- $C = \{ m \mid m$ *is a molecule*$\}$,
- $X \in Con \iff \bigcup X$ *satisfies conditions 1 and 2 of stable approximable mappings,*
- $\{ m \} \vdash m' \iff m' \subseteq m.$

One can regard a molecule as a proposition which is a conjunction of specifications about minimal information.

Proposition 7.7 *The function space construction given in the previous definition is an exponentiation in the category of prime information systems and stable approximable mappings.*

Proof All we have to do is to show that $\mid [\underline{A} \to \underline{B}] \mid$ is isomorphic to the set of stable approximable mappings from $\underline{A}$ to $\underline{B}$. Function space is defined by taking special kinds of stable approximable mappings, the molecules, as propositions. It can be easily checked that by sending each element x in the function space to the union $\bigcup x$ (which is a stable approximable mapping) of molecules in x, one indeed gets an isomorphism.

□

Theorem 7.4 *Lifting is a continuous function* $(\ )_\uparrow : \mathbf{CPO_{pis}} \to \mathbf{CPO_{pis}}$. *Sum, product and function space* $+, \times, \to : \mathbf{CPO^2_{pis}} \to \mathbf{CPO_{pis}}$ *are also continuous functions.*

Proof We take the construction of function space as an example. Other cases are much simpler, hence omitted.

First we check that $\rightarrow$ preserves prime information systems. Let $\underline{A}$, $\underline{B}$ be prime information systems. It is easy to see that $[\underline{A} \rightarrow \underline{B}]$ is an information system. By Definition 7.1, Axiom 6 holds. Suppose a is a molecule. We want to show that $\{ b \mid \{ a \} \vdash b \;\&\; b$ is a molecule $\}$ is finite. But $\{ a \} \vdash b$ if $b \subseteq a$ and a, b are finite sets. Hence $\{ b \mid \{ a \} \vdash b \;\&\; b$ is a molecule $\}$ is finite.

$\rightarrow$ is monotonic in its first argument. Suppose $\underline{A} \trianglelefteq \underline{A}'$. Write

$$\underline{C} = (C, Con, \vdash) = [\underline{A} \rightarrow \underline{B}]$$

and

$$\underline{C}' = (C', Con', \vdash') = [\underline{A}' \rightarrow \underline{B}].$$

We check 1, 2 and 3 in Definition 7.5, to show that $\underline{C} \trianglelefteq \underline{C}'$. Axiom 1 is trivial. Axiom 2. Suppose $\{ c_i \mid 1 \leq i \leq n \} \in Con$. Then clearly $\{ c_i \mid 1 \leq i \leq n \} \subseteq C$ and $\{ c_i \mid 1 \leq i \leq n \} \in Con'$. On the other hand, suppose $\{ c_i \mid 1 \leq i \leq n \} \subseteq C$ and $\{ c_i \mid 1 \leq i \leq n \} \in Con'$. We have

$$\begin{aligned} &\forall S \subseteq \textstyle\bigcup_{1 \leq i \leq n} c_i .\ \bigcup \pi_1 S \in Con_{\underline{A}'} \Longrightarrow \bigcup \pi_2 S \in Con_{\underline{B}} \;\& \\ &\forall \alpha, \beta \in \textstyle\bigcup_{1 \leq i \leq n} c_i .\ \{ \pi_2 \alpha \} = \{ \pi_2 \beta \} \& \ \pi_1 \alpha \cup \pi_1 \beta \in \mid \underline{A}' \mid \\ &\qquad\qquad\qquad\qquad \Longrightarrow \pi_1 \alpha = \pi_1 \beta. \end{aligned}$$

However,

$$\bigcup \pi_1 S \in Con_{\underline{A}} \Longrightarrow \bigcup \pi_1 S \in Con_{\underline{A}'}$$

as $\underline{A} \trianglelefteq \underline{A}'$. Therefore $\{ c_i \mid 1 \leq i \leq n \} \in Con$.

Axiom 3 follows from a similar argument used in showing Axiom 2. Let

$$\underline{A}_0 \trianglelefteq \underline{A}_1 \trianglelefteq \cdots \trianglelefteq \underline{A}_i \trianglelefteq \cdots$$

be a chain of prime information systems. Let m be a molecule of

$$[(\bigsqcup_{i \in \omega} \underline{A}_i) \rightarrow \underline{B}].$$

Then $\bigcup \pi_1 m \Subset \bigcup_{i \in \omega} A_i$. Hence $\bigcup \pi_1 m \subseteq A_j$ for some j, which means m is a molecule of $[\underline{A}_j \rightarrow \underline{B}]$. Thus m is a molecule of $\bigsqcup_{i \in \omega} [\underline{A}_i \rightarrow \underline{B}]$. Therefore $\rightarrow$ is continuous in its first argument. By a similar but easier proof we get that $\rightarrow$ is continuous in its second argument hence it is continuous.

□

At this stage, the reader may wonder why is the construction of function space so different from the one on information systems; Why can't we use propositions of the form (X, Y) or even (X, b) for the function space?

Information systems describe the consistency and the entailment relation on propositions. The entailment is global: Once $X \vdash a$, it holds for the information system irrespective of the particular computation of the type. As the stable approximable mapping suggests, a pair (a, p) should read: The set of propositions a entails the proposition p, and a is a weakest such set. If we take (a, p) as the basic unit of information for the function space, it may lack the global property. Consider the function space on the simple information system $(\{1, 2\}, Con, \vdash)$, where Con is generated by requiring 1, 2 to be consistent and $\vdash$ by $\{2\} \vdash 1$. If we know that x is a computation which produces 2 with the minimal information $\{1, 2\}$, written as $(\{1, 2\}, 2) \in x$, we know that 1 is somehow also produced, since we have $\{2\} \vdash 1$. We can then ask what is the minimal information needed for x to produce 1. There are three possibilities: $(\{1, 2\}, 1) \in x$, $(\{1\}, 1) \in x$, and $(\emptyset, 1) \in x$. Therefore $(\{1, 2\}, 2)$ entails $(\{1, 2\}, 1)$, or $(\{1\}, 1)$, or $(\emptyset, 1)$, but not all of them at the same time since they are inconsistent. This illustrates why we cannot get a global entailment by using propositions of the form (a, p) for the function space.

Our construction of function space works for the example in the following way. There are altogether nine molecules, four containing $(\emptyset, 1)$, three containing $(\{1\}, 1)$, two containing $(\{1, 2\}, 1)$. For example, $\{(\emptyset, 1)\}$ is one of the molecule. Clearly, these nine molecules correspond to the nine complete primes in $[|\underline{A}| \rightarrow_s |\underline{A}|]$, where $\underline{A}$ is the prime information system under consideration.

7.5 Coherent Information Systems

There is a special class of prime information systems for which one can indeed use (X, b) as propositions for the function space. They are the prime information systems with a trivial entailment relation: $X \vdash a$ if and only if $a \in X$. In fact if we require further that Con be binary, in the sense that $X \in Con$ if and only if $\forall a, b \in X.\ \{a, b\} \in Con$, and $\{a\} \approx \{b\}$ implies $a = b$, then they are exactly coherent spaces.

Thus there is no doubt that the theory developed in the previous sections applies to coherent spaces. However, there is one thing we should check: we have to make sure that the solutions to recursive equations in coherent spaces are themselves coherent! Although this is not difficult, we still have to explain why it works.

Since for coherent spaces the consistency predicate *Con* is determined by a conflict relation $\#$, a pair $(E, \#)$ determines a coherent space (here we choose to insist that $\#$ be irreflexive) since the entailment is trivial. Call a structure $(E, \#)$ with $\#$ a symmetric, irreflexive relation on E a *coherent information system.* This name, we admit, does not imply anything profound: coherent information systems are exactly coherent spaces, though presented in a slightly different way. A different presentation is used here since by viewing coherent spaces as a special kind of prime information systems we can directly apply the method developed in the previous sections for recursively defined systems.

There is some technical advantage to work with structures $(E, \mathbin{\#\!\!\!\smile})$, derived from $(E, \#)$ by taking $\mathbin{\#\!\!\!\smile}$ to be $\# \cup \mathbf{1}$, with $\mathbf{1}$ the identity relation. However one can easily recover $(E, \#)$ from $(E, \mathbin{\#\!\!\!\smile})$ by taking $\#$ to be $\mathbin{\#\!\!\!\smile} \setminus \mathbf{1}$.

The rest of the section is going to be brief. We introduce the necessary definitions and leave the properties for the readers to check. There are the following constructions on coherent information systems.

Sum. Let $\underline{E}_1 = (E_1, \mathbin{\#\!\!\!\smile}_1)$, $\underline{E}_2 = (E_2, \mathbin{\#\!\!\!\smile}_2)$ be coherent information systems. Their sum, $\underline{E}_1 + \underline{E}_2$, is a structure $\underline{E} = (E, \mathbin{\#\!\!\!\smile})$ where

$$E = \{1\} \times E_1 \cup \{2\} \times E_2,$$
$$(i, e_1)\,\#\,(j, e_2) \Longleftrightarrow i \neq j \text{ or } (i = j \;\&\; e_1 \,\#_i\, e_2).$$

Product. Let $\underline{E}_1 = (E_1, \mathbin{\#\!\!\!\smile}_1)$, $\underline{E}_2 = (E_2, \mathbin{\#\!\!\!\smile}_2)$ be coherent information systems. Their product, $\underline{E}_1 \times \underline{E}_2$, is a structure $\underline{E} = (E, \mathbin{\#\!\!\!\smile})$ where

$$E = \{1\} \times E_1 \cup \{2\} \times E_2,$$
$$(i, e_1)\,\mathbin{\#\!\!\!\smile}\,(j, e_2) \Longleftrightarrow i = j \;\&\; e_1 \mathbin{\#\!\!\!\smile}_i e_2.$$

Tensor product. Let $\underline{E}_1 = (E_1, \mathbin{\#\!\!\!\smile}_1)$, $\underline{E}_2 = (E_2, \mathbin{\#\!\!\!\smile}_2)$ be coherent information systems. Their tensor product, $\underline{E}_1 \otimes \underline{E}_2$, is a structure

$$\underline{E} = (E, \mathbin{\#\!\!\!\smile})$$

where

$$E = E_1 \times E_2,$$
$$(e_1, e_2) \# (e_1', e_2') \Longleftrightarrow e_1 \#_1 e_1' \text{ or } e_2 \#_2 e_2'.$$

Linear function space. Let $\underline{E}_1 = (E_1, \asymp_1)$, $\underline{E}_2 = (E_2, \asymp_2)$ be coherent information systems. The linear function space, $\underline{E}_1 \multimap \underline{E}_2$, is a structure $\underline{E} = (E, \asymp)$ where

$$E = E_1 \times E_2,$$
$$(e_1, e_2) \asymp (e_1', e_2') \Longleftrightarrow \neg(e_1 \#_1 e_1') \;\&\; e_2 \asymp_2 e_2'.$$

Exponential. Let $\underline{E}_1 = (E_1, \asymp_1)$ be a coherent information system. Its exponential, $!\underline{E}_1$, is a structure $\underline{E} = (E, \asymp)$ where

$$E = \mid \underline{E}_1 \mid^0$$
$$a \# a' \Longleftrightarrow \exists e \in a \, \exists e' \in a' \,.\, e \#_1 e'.$$

Note that there is no lifting construction for coherent information systems. Coherent spaces are not closed under lifting.

The theorem below asserts that all the constructions introduced above preserve coherent information systems, and they induce continuous functions. This implies that it is possible to give meanings to recursively defined coherent information systems involving these constructions. Note that restricted to coherent information systems Definition 7.5 provides another cpo which is written as $\mathbf{CPO}_{\mathrm{cis}}$.

Theorem 7.5 *Exponential is a continuous function* $! : \mathbf{CPO}_{\mathrm{cis}} \to \mathbf{CPO}_{\mathrm{cis}}$. *Sum, product, tensor product, and linear function space*

$$+, \times, \otimes, \multimap \; : \mathbf{CPO}_{\mathrm{cis}}^2 \to \mathbf{CPO}_{\mathrm{cis}}$$

are also continuous functions.

Proof We leave it to the reader to prove that all the constructions give well-defined functions. We prove that exponential and linear function space are continuous. The proof for the rest of the constructions are similar.

To show exponential is monotonic let

$$\underline{E}_1 = (E_1, \asymp_1),$$
$$\underline{E}_2 = (E_2, \asymp_2)$$

be coherent information systems such that $\underline{E}_1 \trianglelefteq \underline{E}_2$. Let $!\underline{E}_1 = (F, \mathbb{W})$ and $!\underline{E}_2 = (F', \mathbb{W}')$. Clearly, then, $F \subseteq F'$. By definition, $a \# b$ if and only if there are $e \in a$, $f \in b$ such that $e \#_1 f$; $a' \#' b'$ if and only if there are $e' \in a'$, $f' \in b'$ such that $e' \#_2 f'$. Thus $a \# b$ if and only if $a, b \in F$ and $a \#' b$ since $\underline{E}_1 \trianglelefteq \underline{E}_2$. Therefore $!\underline{E}_1 \trianglelefteq !\underline{E}_2$.

Now let

$$\underline{E}_1 \trianglelefteq \underline{E}_2 \cdots \trianglelefteq \underline{E}_i \trianglelefteq \cdots$$

be a chain of coherent information systems. Suppose a is a proposition of $!(\bigcup_{i\geq 1} \underline{E}_i)$. We have $a \Subset \bigcup_{i\geq 1} E_i$. Thus $a \Subset E_n$ for some n since a is a finite set. Therefore a is an proposition of $\bigcup_{i\geq 1} !(\underline{E}_i)$. This means ! is continuous on propositions. Thus we have proved that ! is continuous.

The linear function space, $\multimap$, is monotonic: let

$$\underline{E}_1 = (E_1, \mathbb{W}_1),$$
$$\underline{E}_2 = (E_2, \mathbb{W}_2),$$
$$\underline{E}_3 = (E_3, \mathbb{W}_3)$$

be coherent information systems such that $\underline{E}_1 \trianglelefteq \underline{E}_2$. Write $(E, \mathbb{W})$ for $\underline{E}_1 \multimap \underline{E}_3$ and $(E', \mathbb{W}')$ for $\underline{E}_2 \multimap \underline{E}_3$. Clearly $E \subseteq E'$. By definition, $(e_1, e_2)\, \mathbb{W}\, (e_1', e_2')$ if and only if $\neg(e_1 \#_1 e_1')$ and $e_2\, \mathbb{W}_2\, e_2'$; $(e_2, e_2)\, \mathbb{W}'\, (e_2', e_2')$ if and only if $\neg(e_2 \#_2 e_2')$ and $e_2\, \mathbb{W}_2\, e_2'$. Thus

$$(e_1, e_2)\, \mathbb{W}\, (e_1^t, e_2')$$

if and only if $(e_1, e_2), (e_1', e_2') \in E$ and $(e_1, e_2)\, \mathbb{W}'\, (e_1', e_2')$ since $\underline{E}_1 \trianglelefteq \underline{E}_2$. Therefore

$$(\underline{E}_1 \multimap \underline{E}_3) \trianglelefteq (\underline{E}_2 \multimap \underline{E}_3).$$

Similarly $\multimap$ is monotonic in its second argument.

Let

$$\underline{E}_1 \trianglelefteq \underline{E}_2 \cdots \trianglelefteq \underline{E}_i \trianglelefteq \cdots$$

be a chain of coherent information systems. Suppose (e_1, e) is a proposition of $(\bigcup_{i\geq 1} \underline{E}_i) \multimap \underline{E}$, with $\underline{E} = (E, \mathbb{W})$. We have

$$(e_1, e) \in (\bigcup_{i\geq 1} E_i) \times E.$$

Thus $(e_1, e) \in E_n \times E$ for some n. Therefore (e_1, e) is a proposition of $\bigcup_{i\geq 1}(\underline{E}_i \multimap \underline{E})$. This means $\multimap$ is continuous on propositions in its first argument. Similarly it is continuous on propositions in the second argument.

Therefore $\multimap$ is continuous on both its first and second argument, and hence it is continuous.

□

Note that since we use concrete structures of sets to solve equations of coherent information systems the usual domain isomorphism becomes an equality.

Chapter 8

Stable Neighborhoods

Scott topology plays an essential role for the logic of domains developed in Part I. One of the reasons for this is that Scott open sets characterize continuous functions. Scott open sets on stable domains, however, do not characterize stable functions. They do not characterize the stable order, in particular. This makes key rules like

$$\frac{A \leq A' \qquad B' \leq B}{A' \to B' \leq A \to B}$$

inappropriate for the stable case.

The purpose of this chapter is to explore the topological properties related to stable functions and to study the constructions on the resulting topological spaces. We introduce ***stable neighborhoods***, which are a certain kind of open sets, to characterize stable functions. A dI-domain can be seen as a collection of computations of certain type. The stable neighborhoods of the dI-domain can be taken as properties about the computations. Constructions on dI-domains can be seen as ways to combine computations. The corresponding constructions on stable neighborhoods generate proof rules for the logical framework of stable domains.

The organization of this chapter is as follows. Section 8.1 introduces stable neighborhoods and studies their basic properties. Section 8.2 deals with the constructions of stable neighborhoods in the category of stable domains. Section 8.3 deals with the constructions of stable neighborhoods with respect to constructions in the category of coherent spaces. Section 8.4 studies the constructions of stable neighborhoods with respect to the categories for concurrency.

8.1 Stable Neighborhoods

Let us first try to find out what kind of properties we expect the 'open sets' to have in order to characterize stable functions. This is best done by first recalling some of the properties that Scott open sets have. They include

1. A function is continuous if and only if
   the inverse image of an open set is open,
2. For continuous functions f and g, $f \sqsubseteq g$ if and only if
   $f^{-1}(O) \subseteq g^{-1}(O)$ for all open set O,
3. Scott open sets form a topology.

The first thing we notice is that the first property and the third property cannot hold at the same time for stable functions. We illustrate that if a class of open sets of a stable domain has the property that *a function is stable if and only if for any set in this class, the inverse image of the set is still in this class*, it does not necessarily form a topology.

Consider the stable functions from $\mathcal{O}^2$ to $\mathcal{O}$.

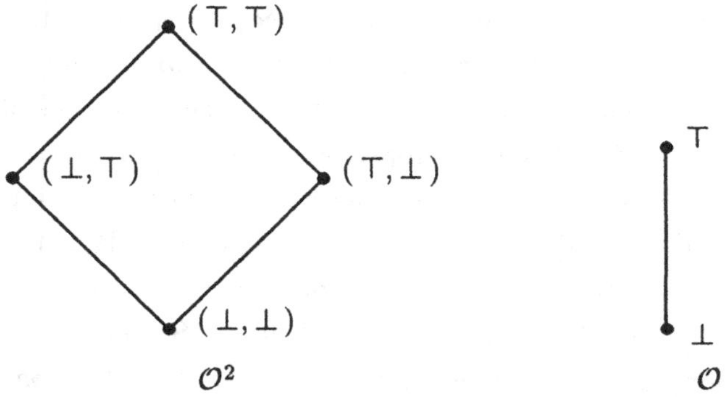

Suppose there were such a topology on both $\mathcal{O}$ and $\mathcal{O}^2$. Then the topology on $\mathcal{O}$ must contain $\{\top\}$ as an open set. This is because otherwise one may get non-monotonic functions. Now take the open set $\{\top\}$ as the starting point. For any stable function f the inverse image $f^{-1}(\{\top\})$ should be an open set in the supposed topology. It is easy to find stable functions so that the inverse images $f^{-1}(\{\top\})$ include $\{(\top, \top)\}$ and the whole domain $\mathcal{O}^2$. The inverse image $f^{-1}(\{\top\})$ of the stable function

$$(\top, \bot) \longmapsto \top, \quad (\bot, \top) \longmapsto \bot$$

is $\{(\top, \top), (\top, \bot)\}$, and the inverse image $f^{-1}(\{\top\})$ of the stable function

$$(\bot, \top) \longmapsto \top, \quad (\top, \bot) \longmapsto \bot$$

is $\{(\top, \top), (\bot, \top)\}$. The existence of the kind of topology on $\mathcal{O}^2$ implies the union of these sets,

$$\{(\top, \top), (\bot, \top), (\top, \bot)\},$$

would be again an open set.

By assumption, to test if a function is stable or not all one has to do is to see if the inverse images of open sets belong to this topology. However the analysis given above shows that the topology on $\mathcal{O}^2$ would coincide with the Scott topology, which allows the non-stable 'parallel-or':

$$\begin{aligned}(\bot, \top) &\longmapsto \top,\\ (\top, \bot) &\longmapsto \top,\\ (\bot, \bot) &\longmapsto \bot.\end{aligned}$$

Therefore the required topology to characterize stable functions does not exist on $\mathcal{O}^2$.

The next best thing to do is to look for a class of open sets which characterizes stable functions in the slightly modified sense:

- a function is stable if and only if the inverse image of a set in this class is still a set in the class,
- for stable functions f, g, $f \sqsubseteq_s g$ if and only if $f^{-1}(O) \sqsubseteq g^{-1}(O)$ for all set O in this class, where $\sqsubseteq$ is a suitable order.

To get some hint let us examine Scott continuous functions from another stand-point. For a Scott domain D, we have

$$[D \to \mathcal{O}] \cong \Omega(D),$$

i.e., the continuous functions $[D \to \mathcal{O}]$ with the pointwise order are isomorphic to Scott open sets of D. The isomorphism is given by $f \longmapsto f^{-1}\{\top\}$.

Now consider a dI-domain D and a stable function $f : D \to \mathcal{O}$. $f^{-1}\{\top\}$ is a Scott-open set as f is continuous. If $x \uparrow y$ and $x, y \in f^{-1}\{\top\}$ then $x \sqcap y \in f^{-1}\{\top\}$ since a stable function preserves greatest lower bounds of compatible elements. This simple analysis leads to

Definition 8.1 *Let D be a dI-domain. U is a stable neighborhood of D if*

- *U is Scott-open, and*
- *$(x \uparrow y \;\&\; x, y \in U) \Longrightarrow x \sqcap y \in U$.*

Write the set of stable neighborhoods of a dI-domain D as $\mathbf{SN}(D)$. $\mathbf{SN}(D)$ does not necessarily form a topology as anticipated. It is closed under finite intersections but not arbitrary unions.

Example 8.1 *In $\mathcal{O}^2$, $\{(\bot, \top), (\top, \top)\}$ and $\{(\top, \bot), (\top, \top)\}$ are stable neighborhoods, but not their union.*

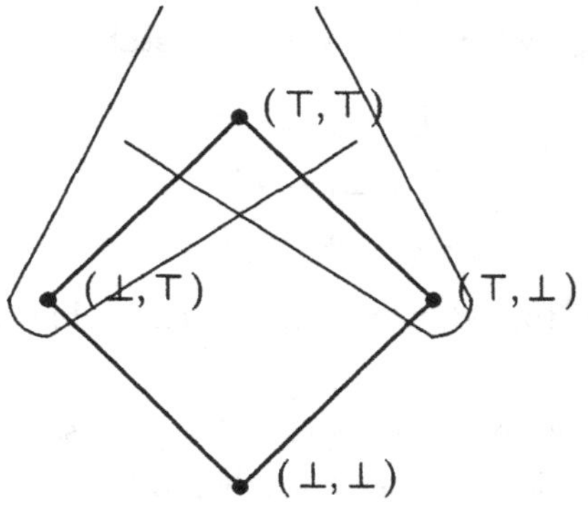

However, stable neighborhoods are closed under *disjoint* unions and finite intersections. To cope with this phenomenon some new terminologies are introduced.

Definition 8.2 *A disjunctive space is a pair $(S, \mathcal{N})$ where S is a set and $\mathcal{N}$ is a collection of subsets of S satisfying the following properties:*

- *$\mathcal{N}$ is closed under finite intersection,*
- *$\mathcal{N}$ is closed under disjoint union, i.e.*
  $(\forall i \in I.\, U_i \in \mathcal{N}) \;\&\; (U_i \cap U_j = \emptyset$ for every $i \neq j) \Longrightarrow \bigcup_{i \in I} U_i \in \mathcal{N}$.

A member of $\mathcal{N}$ is called a neighborhood.

Clearly a topological space is a disjunctive space, but not *vice versa.* However, many useful concepts on topological spaces can be introduced to disjunctive spaces without much change. A *base* of a disjunctive space $\mathcal{N}$ is a subset $\mathcal{B} \subseteq \mathcal{N}$ such that every neighborhood is a disjoint union of members from $\mathcal{B}$. A *sub-base* of $\mathcal{N}$ is a subset $\mathcal{A} \subseteq \mathcal{N}$ such that every neighborhood is a disjoint union of finite intersections of elements of $\mathcal{A}$.

Suppose $x \in S$ and $\mathcal{N}$ is a disjunctive space on S. Let $\mathcal{P}_x$ be the collection of neighborhoods each of which contains x. Then $\mathcal{P}_x$ has the following properties:

1. $U \in \mathcal{P}_x \;\&\; U \subseteq V \in \mathcal{N} \implies V \in \mathcal{P}_x$,
2. $U \in \mathcal{P}_x \;\&\; V \in \mathcal{P}_x \implies U \cap V \in \mathcal{P}_x$, and
3. for any pairwise disjoint collection of U_i's,
   $\bigcup_{i \in I} U_i \in \mathcal{P}_x \implies U_i \in \mathcal{P}_x$ for some i.

A space is called T_0 if whenever $\mathcal{P}_x = \mathcal{P}_y$, we also have $x = y$. Forgetting about the point x, we call any collection $\mathcal{P}$ of neighborhoods with the foregoing three properties a *complete prime filter* in $\mathcal{N}$. A space is *sober* if it is T_0 and for each complete prime filter $\mathcal{P}$ in $\mathcal{N}$, there is some point x such that $\mathcal{P}_x = \mathcal{P}$. Intuitively a space is sober if it is completely determined by its lattice of properties (neighborhoods).

The stable neighborhoods of a dI-domain indeed form a disjunctive space.

Proposition 8.1 *Let D be a dI-domain. Then the collection of the stable neighborhoods of D is a disjunctive space. Moreover,* $(\mathbf{SN}(D), \subseteq)$ *is a lattice.*

Proof That $(D,\ \mathbf{SN}(D))$ is a disjunctive space is easy to check.

Let $U, V \in \mathbf{SN}(D)$. Clearly $U \cap V$ is Scott-open. $x \uparrow y \;\&\; x, y \in U \cap V$ implies $x \sqcap y \in U \;\&\; x \sqcap y \in V$. Therefore $U \cap V \in \mathbf{SN}(D)$, *i.e.*, $U \sqcap V = U \cap V$. To show that $\mathbf{SN}(D)$ has joins, we first introduce a binary operation $\diamond$ between two open sets A and B:

$$A \diamond B =_{def} \{ d \mid \exists x, y \in A \cup B.\, (x \uparrow y \;\&\; d \sqsupseteq x \sqcap y)\}.$$

It is easy to see that A and B open implies $A \diamond B$ open. Now for $U, V \in \mathbf{SN}(D)$, let

$$K_0 = U \diamond V, \quad K_1 = K_0 \diamond K_0, \quad \dots \quad K_n = K_{n-1} \diamond K_{n-1}, \dots$$

We claim that $\bigcup_{i\in\omega} K_i = U \sqcup V$. It is enough to check that $\bigcup_{i\in\omega} K_i$ is a stable neighborhood, and it actually is:

$$\begin{aligned} d \uparrow d' \ \& \ d, d' \in \bigcup_{i\in\omega} K_i &\Longrightarrow \exists m, n \in \omega.\ d \in K_n \ \& \ d' \in K_m \\ &\Longrightarrow d \in K_{n+m} \ \& \ d' \in K_{n+m} \\ &\Longrightarrow d \sqcap d' \in K_{n+m+1} \subseteq \bigcup_{i\in\omega} K_i. \end{aligned}$$

□

SN(D) is, moreover, a complete lattice. The operation $\diamond$ can be extended to an arbitrary number of stable neighborhoods and $\bigcup K_i$ will be their least upper bound. Then the meet of an arbitrary number of stable neighborhoods is the join of all the stable neighborhoods contained in their intersection.

However, **SN**(D) need not be distributive, as Example 8.2 shows.

Example 8.2 *Consider* $\mathcal{T} \times \mathcal{O}$.

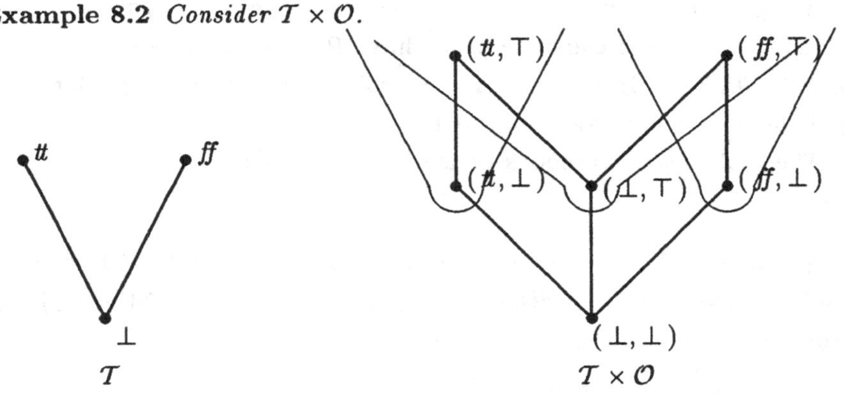

Let

$$\begin{aligned} A &= \{ (tt, \bot), (tt, \top) \}, \\ B &= \{ (\bot, \top), (tt, \top), (ff, \top) \}, \\ C &= \{ (ff, \bot), (ff, \top) \}. \end{aligned}$$

Then $A \sqcap (B \sqcup C) = A$, *but*

$$(A \sqcap B) \sqcup (A \sqcap C) = \{ (tt, \top), (ff, \top) \} \neq A.$$

Note that least upper bounds in $(\mathbf{SN}(D), \subseteq)$ may not be the same as unions, so we write $\sqcup$. The following definition will be used later.

Definition 8.3 *A compact stable neighborhood of* $\mathbf{SN}(D)$ *is a set which is both a stable neighborhood and a compact Scott open set. Write* $\mathbf{KSN}(D)$ *for the set of compact stable neighborhoods of* $\mathbf{SN}(D)$. *A prime stable neighborhood of* $\mathbf{SN}(D)$ *is just a prime open set* P *such that* $P = \uparrow d$ *for some finite element* d *in* D.

When $f : D \to \mathcal{O}$ is a stable function, $f^{-1}(\top)$ is a stable neighborhood. On the other hand, suppose U is a stable neighborhood of D. Then $F(U)$ is a stable function, where

$$F(U)(x) = \begin{cases} \top & \text{if } x \in U, \\ \bot & \text{if } x \in (D \setminus U). \end{cases}$$

However, different from Scott continuous functions, set inclusion on stable neighborhoods does not determine the stable order. In $[\mathcal{O} \to_s \mathcal{O}]$, for example, two stable functions $\lambda x.x$ and $\lambda x.\top$ have the property

$$\forall U \in \mathbf{SN}(\mathcal{O}).\ (\lambda x.x)^{-1}(U) \subseteq (\lambda x.\top)^{-1}(U),$$

but we do not have $\lambda x.x \sqsubseteq_s \lambda x.\top$.

To find an order on the stable neighborhoods which determines the stable order on functions, suppose U, V are stable neighborhoods of D such that $F(U) \sqsubseteq_s F(V)$, where $F : \mathbf{SN}(D) \to [D \to_s \mathcal{O}]$ is defined in the previous paragraph. Then for any x, y in D, $x \sqsubseteq y$ implies

$$F(U)(x) = F(U)(y) \sqcap F(V)(x).$$

Obviously $x \in V$ when $x \in U$, *i.e.*, $U \subseteq V$. Moreover, whenever $x \sqsubseteq y \in U$ but $x \notin U$, it must also be true that $x \notin V$. This means a minimal point of U must also be a minimal point of V.

Definition 8.4 *Let* D *be a dI-domain. The set of minimal points of a stable neighborhood* U *of* D, *write* μU, *consists of* $m \in U$ *such that*

$$\forall x \sqsubseteq m.\ [x \in U \Rightarrow x = m].$$

For $U, V \in \mathbf{SN}(D)$, U *minimally less than* V, *write* $U \sqsubseteq_\mu V$, *if* $\mu U \subseteq \mu V$.

The following proposition is immediate.

Proposition 8.2 *Let D be a dI-domain. Then there is a 1-1, order preserving correspondence between $[D \rightarrow_s \mathcal{O}]$ and $\mathbf{SN}(D)$, with the stable order on $[D \rightarrow_s \mathcal{O}]$ and $\sqsubseteq_\mu$ on $\mathbf{SN}(D)$. It is given by*

$$f \longmapsto f^{-1}(\{\top\}).$$

Clearly $\sqsubseteq_\mu$ is an equivalence relation. $U \sqsubseteq_\mu V$ implies $U \subseteq V$ but not vice versa. Notice that if $U \sqsubseteq_\mu W$ and $U \subseteq V \subseteq W$, then $U \sqsubseteq_\mu V$. $U \sqsubseteq_\mu V$ if and only if $\exists W \in \mathbf{SN}(D).\ V = U \cup W\ \&\ U \cap W = \emptyset$. Every minimal point of a stable neighborhood is a finite element.

Proposition 8.3

$$(A \sqsubseteq_\mu B\ \&\ S \sqsubseteq_\mu T) \implies (A \cap S) \sqsubseteq_\mu (B \cap T)$$

where $A, B, S, T \in \mathbf{SN}(D)$.

Proof We have, for some A', $S' \in \mathbf{SN}(D)$, $B = A \cup A'$, $A \cap A' = \emptyset$, $T = S \cup S'$, $S \cap S' = \emptyset$. It follows that

$$B \cap T = (A \cap S) \cup \Big[(A \cap S') \cup (A' \cap S) \cup (A' \cap S')\Big],$$

where

$$(A \cap S) \cap \Big[(A \cap S') \cup (A' \cap S) \cup (A' \cap S')\Big] = \emptyset$$

and

$$(A \cap S') \cup (A' \cap S) \cup (A' \cap S') \in \mathbf{SN}(D).$$

Therefore $(A \cap S) \sqsubseteq_\mu (B \cap T)$.

□

From the properties of stable neighborhood one can easily derive

Proposition 8.4 *$U \in \mathbf{SN}(D)$ implies there is some $K \subseteq D^0$, a pairwise incompatible set, such that*

$$U = \uparrow K.$$

Stable neighborhoods indeed characterize stable functions. Suppose f from D to E is stable. For any $g : E \rightarrow \mathcal{O}$, there is a unique stable function $h : D \rightarrow \mathcal{O}$ which makes the following diagram commute.

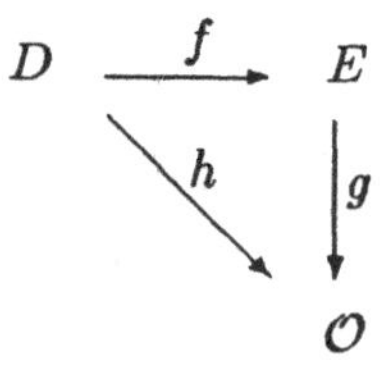

This implies for any $g : E \to \mathcal{O}$, $f^{-1}(g^{-1}(\top)) = h^{-1}(\top) \in \mathbf{SN}(D)$, or, for any $U \in \mathbf{SN}(E)$, $f^{-1}(U) \in \mathbf{SN}(D)$. In general, we have

Theorem 8.1 *Let D, E be dI-domains. $f : D \to E$ is stable if and only if*

$$\forall U \in \mathbf{SN}(E).\ f^{-1}(U) \in \mathbf{SN}(D).$$

Proof Only if: As shown above.

If: Such f is monotonic: Let $d, d' \in D$ and $d' \sqsubseteq d$. For any finite e in E, if $e \sqsubseteq f(d')$ then $d' \in f^{-1}(\uparrow e)$. $f^{-1}(\uparrow e) \in \mathbf{SN}(D)$ as $\uparrow e \in \mathbf{SN}(E)$. So $d \in f^{-1}(\uparrow e)$, *i.e.*, $f(d) \sqsupseteq e$. Therefore $f(d') \sqsubseteq f(d)$.

f preserves directed sups. Assume X is a directed set of D. $e \sqsubseteq f(\bigsqcup X)$ implies $\bigsqcup X \in f^{-1}(\uparrow e)$, which is open. Hence $\exists x \in X$ such that

$$x \in f^{-1}(\uparrow e),$$

i.e., $f(x) \sqsupseteq e$. Therefore $f(\bigsqcup X) \sqsubseteq \bigsqcup f(X)$. Thus f is continuous.

To see f is stable notice we have, for any $e' \in E^0$,

$$\begin{aligned} & x \uparrow y \ \&\ x, y \in D \ \&\ e' \sqsubseteq f(x) \sqcap f(y) \\ & \quad \Longrightarrow\ x \in f^{-1}(\uparrow e') \ \&\ y \in f^{-1}(\uparrow e') \\ & \quad \Longrightarrow\ x \sqcap y \in f^{-1}(\uparrow e') \\ & \quad \Longrightarrow f(x \sqcap y) \sqsupseteq e'. \end{aligned}$$

Therefore $f(x \sqcap y) = f(x) \sqcap f(y)$.

□

Similarly to Theorem 8.1 one can show that $f : D \to E$ is linear, stable if and only if $\forall U \in \mathbf{PSN}(E).\ f^{-1}(U) \in \mathbf{PSN}(D)$, where $\mathbf{PSN}(E)$ stands for the collection of *very prime* stable neighborhoods of E, those which take the form $\uparrow p$ for some complete prime p of E.

Suppose $f,\ f' : D \to E$ are stable functions. If $f \sqsubseteq_s f'$ then clearly for any $g : E \to \mathcal{O}$, $g \circ f \sqsubseteq_s g \circ f'$. By Proposition 8.2 we have, for any

$g : E \to \mathcal{O}$, $f^{-1}(g^{-1}(\top)) \sqsubseteq_\mu (f')^{-1}(g^{-1}(\top))$, or, for any $U \in \mathbf{SN}(E)$, $f^{-1}(U) \sqsubseteq_\mu (f')^{-1}(U)$. The other way round is also true. We have

Theorem 8.2 *Let f, g be members of $[D \to_s E]$. $f \sqsubseteq_s g$ if and only if*

$$\forall U \in \mathbf{SN}(E).\ f^{-1}(U) \sqsubseteq_\mu g^{-1}(U).$$

Proof It is enough to prove sufficiency. For any $x \in D$ and $d \sqsubseteq f(x)$ in E^0, We have $\uparrow d \in \mathbf{SN}(E)$ and $f^{-1}(\uparrow d) \subseteq g^{-1}(\uparrow d)$. Therefore $d \sqsubseteq g(x)$ and hence $f \sqsubseteq g$. Let $x \sqsubseteq y$ in D, $d' \in E^0$, and $y \sqsupseteq y_0 \in \mu f^{-1}(\uparrow d')$. By assumption, $y_0 \in \mu g^{-1}(\uparrow d')$.

$$\begin{aligned} d' \sqsubseteq f(y) \sqcap g(x) &\Longrightarrow x \in g^{-1}(\uparrow d')\ \&\ y \in f^{-1}(\uparrow d') \\ &\Longrightarrow y_0 \uparrow x\ \&\ y_0 \sqcap x \in g^{-1}(\uparrow d') \\ &\Longrightarrow y_0 \sqcap x = y_0 \quad (\text{ as } y_0 \in \mu g^{-1}(\uparrow d')\ ) \\ &\Longrightarrow x \sqsupseteq y_0 \\ &\Longrightarrow f(x) \sqsupseteq f(y_0) \sqsupseteq d'. \end{aligned}$$

Therefore $f(x) \sqsupseteq f(y) \sqcap g(x)$. In other words, $f \sqsubseteq_s g$.

□

8.2 Constructions in DI

This section studies constructions of stable neighborhoods in **DI** with respect to constructions in this category. As pointed out earlier, a dI-domain can be seen as a collection of computations of certain type. The stable neighborhoods of the dI-domain can be taken as properties about the computations. Constructions on dI-domains can be seen as ways to combine computations. Now suppose x is a computation of type D, having property A, written $x \models A$ and y is a computation of type E, having property B, written $y \models B$. If we combine a computation x of D with y of E to get a computation $(x \text{ op } y)$ of type $[D \text{ op } E]$ (Here op is some domain construction like sum, product, or stable function space), can we deduce some property of $(x \text{ op } y)$ from the facts $x \models A$ and $y \models B$? This question leads to constructions $(A \text{ op } B)$ on properties A and B so that from $x \models A$ and $y \models B$ one deduces $(x \text{ op } y) \models (A \text{ op } B)$.

There can be different ways to combine a stable neighborhood A of D with a stable neighborhood B of E to get a stable neighborhood $(A \text{ op } B)$

of $[D \text{ op } E]$. But the following are some criteria for good constructions. Here we are primarily interested in the behaviors of the constructions on compact stable neighborhoods.

- if A is a stable neighborhood of D and B a stable neighborhood of E, $(A \text{ op } B)$ should be a stable neighborhood of $[D \text{ op } E]$,
- stable neighborhoods of the form $(A \text{ op } B)$ should form a subbase of ($[D \text{ op } E]$, $\mathbf{SN}([D \text{ op } E])$),
- if (D, $\mathbf{SN}(D)$) and (E, $\mathbf{SN}(E)$) are sober spaces then so is ($[D \text{ op } E]$, $\mathbf{SN}([D \text{ op } E])$).

The first condition requires that $(A \text{ op } B)$ is well-defined. The second condition states that stable neighborhoods of the form $(A \text{ op } B)$ are expressive enough. The third condition implies that we can reconstruct the points of the domain from the collection of stable neighborhoods. We claim that *all the constructions to be introduced do have the above properties except the* $\leftrightarrow_{ps}$ *in the category* $\mathbf{SEV}^*_{syn}$. However some of the proofs are obvious and we do not always explicitly check them all.

Suppose D and E are dI-domains. Similar to Scott domains, it is easy to get the stable neighborhoods of $D + E$, $D \times E$ and $D_\perp$ from those of D and E with the requirements for good constructions satisfied. The question is: how can we get the stable neighborhoods of $[D \to_s E]$ from those of D and E directly?

Let us be reminded of how we dealt with this problem for the Scott topology. If A and B are compact Scott open sets of D and E, respectively, then $A \to B = \{ h : D \to E \mid A \subseteq h^{-1}(B) \}$ is a compact Scott open set of $[D \to E]$ (Proposition 2.8). There is another way to look at this. Since A and B are Scott open, they correspond to some functions $f_A : D \to \mathcal{O}$, $g_B : E \to \mathcal{O}$. Set inclusion on open sets determines the pointwise order. Hence $h \in A \to B$ if and only if $f_A \sqsubseteq g_B \circ h$ (see the diagram below).

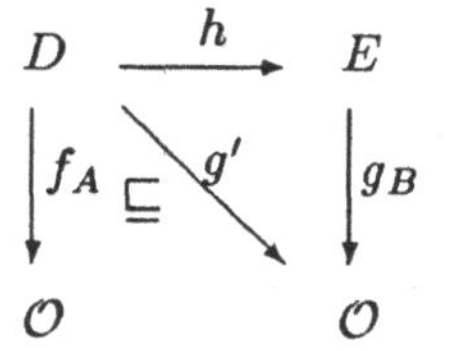

This suggests that for dI-domains we should use the diagram

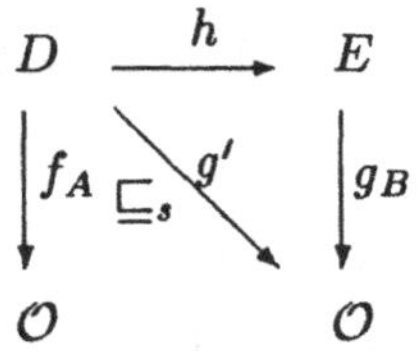

where the pointwise order is replaced by the stable order. Thus one should define $h \in A \to B$ as meaning $f_A \sqsubseteq_s g_B \circ h$. By the analysis given just before Theorem 8.2 however, $f_A \sqsubseteq_s g_B \circ h$ if and only if $A \sqsubseteq_\mu h^{-1}(B)$. That leads to the following definition.

Definition 8.5 *Let D, E be dI-domains, $A \in \mathbf{KSN}(D)$ and $B \in \mathbf{KSN}(E)$. Define $A \to B$ to be the set*

$$\{f \in [D \to_s E] \mid A \sqsubseteq_\mu f^{-1}B\}.$$

We have to show that the construction given in Definition 8.5 does meet the first criteria set at the beginning of the section, that is, the construction is well-defined. But before dealing with that issue let us try to explain intuitively what do we mean for a computation to have a property like $A \to B$ given in the above definition.

Following the view mentioned at the beginning of this section, let us think of a stable function f as a computation of type $[D \to_s E]$ which consumes some information of type D and produces some information of type E (here we can identify the computations of type D and E as data, or information, of type D and E, respectively). What does it mean intuitively for a computation of $[D \to_s E]$ to have a property $A \to B$, where A is a property of type D and B is a property of type E? The properties appropriate for stable functions are those which are determined by a set of incompatible minimal information. We can say that f has property $A \to B$ if f can produce some information with property B from any input information with property A and, moreover, a necessary information of property A is also a necessary information for f to produce some information with property B. We can also say that f has property $A \to B$ if whenever f can produce an output (information) with property B, there is always some minimal input information x for f to do so. If this minimal information x happens to be consistent with property A, then it must also be a minimal information of property A.

Now we have

Theorem 8.3 *Assume* $A \in \mathbf{KSN}(D)$ *and* $B \in \mathbf{KSN}(E)$. *Then*

$$(A \to B) \in \mathbf{KSN}([D \to_s E]).$$

Proof First we prove that $(A \to B)$ is Scott open. It is a direct consequence of Theorem 8.2 that $(A \to B)$ is upwards closed. Suppose

$$f_0 \sqsubseteq_s f_1 \sqsubseteq_s \cdots \sqsubseteq_s f_n \sqsubseteq_s \cdots$$

is a chain in $[D \to_s E]$ and $\bigsqcup_{i \in \omega} f_i \in (A \to B)$. By definition,

$$\mu A \subseteq \mu(\bigsqcup_{i \in \omega} f_i)^{-1} B.$$

As A is compact, μA is finite. Let

$$\mu A = \{a_1, a_2, \cdots, a_m\}.$$

Clearly $a_j \in \mu(\bigsqcup_{i \in \omega} f_i)^{-1} B$ implies $\bigsqcup_{i \in \omega} f_i(a_j) \in B$. For each $1 \leq j \leq m$, there exists I_j such that $f_{I_j}(a_j) \in B$, as B is open. Let n be the biggest of the I_j's. We have $f_n(a_j) \in B$ for all $1 \leq j \leq m$. It is then easy to see that $\mu A \subseteq \mu f_n^{-1} B$, *i.e.* $f_n \in (A \to B)$. Namely, $(A \to B)$ is open.

Assume $f, g \in (A \to B)$ and $f \uparrow g$. By Theorem 8.2 again we have

$$\mu(f \sqcap g)^{-1}(B) \subseteq \mu f^{-1}(B) \cap \mu g^{-1}(B).$$

On the other hand,

$$\begin{aligned} x \in & \mu f^{-1}(B) \cap \mu g^{-1}(B) \\ & \Longrightarrow f(x) \in B \ \&\ g(x) \in B \\ & \Longrightarrow f(x) \sqcap g(x) \in B \\ & \Longrightarrow (f \sqcap g)(x) \in B \\ & \Longrightarrow x \in \mu(f \sqcap g)^{-1}(B). \end{aligned}$$

Hence $\mu(f \sqcap g)^{-1}(B) = \mu f^{-1}(B) \cap \mu g^{-1}(B)$. Now

$$\begin{aligned} f \in (A \to B) \ \&\ g \in (A \to B) & \Longrightarrow A \sqsubseteq_\mu f^{-1}(B) \ \&\ A \sqsubseteq_\mu g^{-1}(B) \\ & \Longrightarrow \mu A \subseteq \mu f^{-1}(B) \cap \mu g^{-1}(B) \\ & \Longrightarrow \mu A \subseteq \mu(f \sqcap g)^{-1}(B). \end{aligned}$$

Therefore $f \sqcap g \in (A \to B)$, and $(A \to B)$ is a stable neighborhood.

To show that $(A \to B)$ is compact we first prove that stable neighborhoods of the form $(\uparrow a \to \uparrow b)$ are compact, where $a \in D^0$, $b \in E^0$. Let $Q_a = \{\, c \in D \mid c \sqsubseteq a \,\}$, $P_b = \{\, p \in E^p \mid p \sqsubseteq b \,\}$, and

$$F_a^b = \{\, f \mid f \text{ is a step function and } \mu f \subseteq Q_a \times P_b \ \&\ a \in \mu f^{-1}(\uparrow b) \,\}.$$

We claim that

$$(\uparrow a \to \uparrow b) = \bigcup_{g \in F_a^b} \uparrow g.$$

It is obvious that

$$(\uparrow a \to \uparrow b) \supseteq \bigcup_{g \in F_a^b} \uparrow g.$$

On the other hand, let $f \in (\uparrow a \to \uparrow b)$. It can be shown that $[\,a,\, b,\, f\,] \in F_a^b$, where

$$[\,a,\, b,\, f\,] =_{\mathrm{def}} \lambda x.\, b \sqcap f(a \sqcap x).$$

Hence $f \in \bigcup_{g \in F_a^b} \uparrow g$ since $f \sqsupseteq_s [\,a,\, b,\, f\,]$. Therefore $(\uparrow a \to \uparrow b)$ is compact since F_a^b is a finite set.

Write $A = \bigcup_{i \in I}(\uparrow a_i)$, $B = \bigcup_{j \in J}(\uparrow b_j)$, where I and J are finite and a_i's are pairwise incompatible, b_j's are pairwise incompatible. It is easy to see that

$$\begin{aligned} A \to B &= (\textstyle\bigcup_{i \in I} \uparrow a_i) \to (\bigcup_{j \in J} \uparrow b_j) \\ &= \textstyle\bigcap_{i \in I}(\uparrow a_i \to \bigcup_{j \in J} \uparrow b_j\,) \\ &= \textstyle\bigcap_{i \in I}[\,\bigcup_{j \in J}(\uparrow a_i \to \uparrow b_j\,)\,]. \end{aligned}$$

Hence $A \to B$ is compact.

□

We have to restrict A and B to compact open sets. Otherwise $(A \to B)$ can be a non-open set, hence not a stable neighborhood. Consider the dI-domain $\mathcal{N}_\perp$. Then set $\mathcal{N}$ is clearly a stable neighborhood of $\mathcal{N}_\perp$. The set, $\{\, f : \mathcal{N}_\perp \to \mathcal{O} \mid \mathcal{N} \subseteq_\mu f^{-1}(\top) \,\}$, is not open since one can easily produce a chain of stable functions whose limit is in this set but not any of the finite approximations.

Unlike the case for the function space of SFP domains,

$$\uparrow a \to \uparrow b$$

need not be a prime stable neighborhood when a and b are finite elements.

Proposition 8.5 *Let* $\{ (a_i, p_i) \mid i \in I \}$ *be a stable joinable set (see Theorem 6.3). Then*

$$f \in (\uparrow a_j \to \uparrow p_j)$$

for every $j \in I$, *where* f *is an abbreviation for the stable function*

$$\bigsqcup_{i \in I} [a_i, p_i]$$

determined by the stable joinable set.

Proof We have

$$\begin{aligned} f(a_j) &= \bigsqcup \{ p_i \mid a_i \sqsubseteq a_j \} \\ &\sqsupseteq p_j . \end{aligned}$$

Hence $a_j \in f^{-1}(p_j \uparrow)$. Let $y \sqsubseteq a_j$ and $f(y) \sqsupseteq p_j$, *i.e.* $\bigsqcup \{ p_i \mid a_i \sqsubseteq y \} \sqsupseteq p_j$. Since p_j is a complete prime, $p_i \sqsupseteq p_j$ for some i with $a_i \sqsubseteq y$. By stable joinable properties, there is some k such that $p_k = p_j$ and $a_k \sqsubseteq a_i$, which implies $a_k = a_j$ since $a_k \uparrow a_j$. Hence $y = a_j$. This means $a_j \in \mu f^{-1}(p_j \uparrow)$.

□

The theorem below implies the second and the third properties required of a construction of the stable neighborhoods set at the beginning of the section. In particular, this theorem says that for a given stable joinable set, if we take the intersection of the stable neighborhoods $(\uparrow a_i \to \uparrow p_i)$, $i \in I$, with I finite, we get a prime stable neighborhood consisting of all the stable functions which dominate the stable function $\bigsqcup_{i \in I} [a_i, p_i]$ under the stable order.

Theorem 8.4 *Let* $\{ (a_i, p_i) \mid i \in I \}$ *be a stable joinable set. Then*

$$\bigcap_{i \in I} (\uparrow a_i \to \uparrow p_i) = \uparrow (\bigsqcup_{i \in I} [a_i, p_i]).$$

Proof We know from Proposition 8.5 that

$$\bigsqcup_{i \in I} [a_i, p_i] \in \bigcap_{i \in I} (\uparrow a_i \to \uparrow p_i).$$

It is enough to show that $\bigsqcup_{i \in I} [a_i, p_i]$ is less than or equal to any other stable function in $\bigcap_{i \in I} (\uparrow a_i \to \uparrow p_i)$. Let g be a stable function in

$$\bigcap_{i \in I} (\uparrow a_i \to \uparrow p_i).$$

For any $i \in I$, $g(a_i) \sqsupseteq p_i$. Therefore for any x in D

$$\bigsqcup\{ p_k \mid a_k \sqsubseteq x \} \sqsubseteq \bigsqcup\{ g(a_k) \mid a_k \sqsubseteq x \} \\ \sqsubseteq g(x),$$

i.e., $\bigsqcup_{i \in I}[a_i, p_i](x) \sqsubseteq g(x)$.

Suppose $x, y \in D$ and $x \sqsubseteq y$. Let $p \sqsubseteq \bigsqcup\{ p_j \sqcap g(x) \mid a_j \sqsubseteq y \}$, where p is a complete prime. $p \sqsubseteq p_j \sqcap g(x)$ for some j. Therefore, there exists s such that $p = p_s$ and $a_j \sqsupseteq a_s$. $g(a_s \sqcap x) = g(a_s) \sqcap g(x) \sqsupseteq p_s$. This implies, as $g \in (\uparrow a_s \to \uparrow p_s)$, $a_s \sqcap x = a_s$, or $a_s \sqsubseteq x$. Hence $p \sqsubseteq \bigsqcup\{ p_i \mid a_i \sqsubseteq x \}$. By the prime algebraicity of E,

$$g(x) \sqcap \bigsqcup\{ p_i \mid a_i \sqsubseteq y \} = \bigsqcup\{ p_j \sqcap g(x) \mid a_j \sqsubseteq y \} \\ \sqsubseteq \bigsqcup\{ p_i \mid a_i \sqsubseteq x \}.$$

Now it is easy to see that $\bigsqcup_{i \in I}[a_i, p_i] \sqsubseteq_s g$.

□

From this theorem we can also see that it is possible to get all the compact stable neighborhoods of $[D \to_s E]$ by finite union and finite intersection of stable neighborhoods of the form $A \to B$, where A, B are compact stable neighborhoods of D and E, respectively. Moreover we can get every prime stable neighborhood in the stable function space in this way. Therefore, we have, for $f, g \in [D \to_s E]$, $f \sqsubseteq_s g$ if and only if $f \in (A \to B)$ implies $g \in (A \to B)$ for all $A \in \mathbf{KSN}(D)$, $B \in \mathbf{KSN}(E)$.

Some proof rules now follow. We let the rules take care of the types themselves.

Proposition 8.6 *Let A, B, C, D be stable neighborhoods and a a finite element. We have*

$$(A \cap B = \emptyset) \Longrightarrow (A \cup B) \to C = (A \to C) \cap (B \to C),$$
$$(A \cap B = \emptyset) \Longrightarrow \uparrow a \to (A \cup B) = (\uparrow a \to A) \cup (\uparrow a \to B),$$
$$(A \to C) \cap (B \to D) \subseteq (A \cap B) \to (C \cap D).$$

Proof Only the last inequality needs verification. Let

$$f \in (A \to C) \cap (B \to D).$$

We have $A \sqsubseteq_\mu f^{-1}C$ and $B \sqsubseteq_\mu f^{-1}D$. One can check that

$$A \cap B \sqsubseteq_\mu (f^{-1}C) \cap (f^{-1}D).$$

But $(f^{-1}C) \cap (f^{-1}D) = f^{-1}(C \cap D)$. Therefore the desired inequality follows.

□

It is important to note however, the rule

$$(A' \subseteq A \ \&\ B \subseteq B') \Longrightarrow (A \to B) \subseteq (A' \to B')$$

which is sound for Scott continuous functions and appeared in Hoare logic, is no longer valid for stable functions. That is because with stable functions, $f \in A \to B$ if and only if $A \sqsubseteq_\mu f^{-1}B$, *not* simply $A \subseteq f^{-1}B$!

Fortunately, the following rule will be helpful.

Proposition 8.7 *Let $a \in D^0$ and $b, c \in E^0$, where D, E are dI-domains.*

$$c \sqsubseteq b \Longrightarrow (\uparrow a \to \uparrow b) \subseteq \bigcup_{a' \sqsubseteq a} (\uparrow a' \to \uparrow c).$$

Note that if $a' \neq a''$ and $a' \uparrow a''$ then $(\uparrow a' \to \uparrow c) \cap (\uparrow a'' \to \uparrow c) = \emptyset$. Hence $\bigcup_{a' \sqsubseteq a} (\uparrow a' \to \uparrow c)$ is a stable neighborhood of $[D \to_s E]$ since it is a disjoint union.

Proof Suppose $f \in (\uparrow a \to \uparrow b)$. Then $a \in \mu f^{-1} \uparrow b$. We have $f(a) \sqsupseteq c$. Let $a'' = \bigsqcap \{ x \mid x \sqsubseteq a \ \&\ f(x) \sqsupseteq c \}$. Clearly $a'' \sqsubseteq a$ and $a'' \in \mu f^{-1} \uparrow c$. Hence $f \in \bigcup_{a' \sqsubseteq a} (\uparrow a' \to \uparrow c)$.

□

8.3 Constructions in $\mathbf{COH}_l$

Since coherent spaces are special kinds of dI-domains, our results in the previous section can be applied to them. However, the category of coherent spaces has some type constructions of its own, which we deal with now. Of course, for each construction we still have to guarantee that the requirements set at the beginning of the previous section are met.

We focus on the stable-neighborhood constructions in $\mathbf{COH}_l$. Once this is clear, we can get the constructions in $\mathbf{COH}_s$ through the exponential construction, which is one of the functors for the adjunction between $\mathbf{COH}_l$ and $\mathbf{COH}_s$.

Now we introduce constructions on the disjunctive spaces related to the tensor product, the linear function space, and the exponential. We show that all these constructions preserve compactness. We also give rules which indicate how those constructions interact with unions and intersections.

Theorem 8.5 *Let $\mathcal{F}_0$, $\mathcal{F}_1$ be coherent spaces and*

$$A \in \mathbf{KSN}(\mathcal{F}_0), \quad B \in \mathbf{KSN}(\mathcal{F}_1).$$

Then $A \otimes B \in \mathbf{KSN}(\mathcal{F}_0 \otimes \mathcal{F}_1)$, where

$$A \otimes B = \{ x \in \mathcal{F}_0 \otimes \mathcal{F}_1 \mid \exists x_0 \in A, x_1 \in B.\ x_0 \times x_1 \Subset x \}.$$

Proof Clearly $A \otimes B$ is a Scott open set. Now assume x, $y \in A \otimes B$ and $x \uparrow y$. Then there exist x_0, $x_1 \in A$, y_0, $y_1 \in B$ such that

$$x_0 \times y_0 \Subset x \quad \text{and } x_1 \times y_1 \Subset y.$$

We have $x_0 \uparrow x_1$ and $y_0 \uparrow y_1$. Therefore $x_0 \cap x_1 \in A$, $y_0 \cap y_1 \in B$. Also, $(x_0 \cap x_1) \times (y_0 \cap y_1) \Subset x \cap y$. Hence $x \cap y \in A \otimes B$.

To check compactness we first show that stable neighborhoods of the form

$$(\uparrow a \otimes \uparrow b)$$

are compact, where a, b are finite configurations of $\mathcal{F}_0$ and $\mathcal{F}_1$, respectively. Clearly

$$a \times b \in (\uparrow a \otimes \uparrow b).$$

On the other hand, suppose

$$u \in (\uparrow a \otimes \uparrow b).$$

Then by definition, there exist a', b' with

$$a \subseteq a' \in \mathcal{F}_0, b \subseteq b' \in \mathcal{F}_1$$

such that $a' \times b' \Subset u$. Hence $a \times b \Subset u$. We have shown that

$$(\uparrow a \otimes \uparrow b) = \uparrow (a \times b).$$

Therefore $(\uparrow a \otimes \uparrow b)$ is compact. By Proposition 8.4 and the fact that

$$A \otimes (B \cup C) = (A \otimes B) \cup (A \otimes C)$$

when $B \cap C = \emptyset$, we deduce that any $A \otimes B$ as specified in Theorem 8.5 is compact.

□

From the proof we can see that

$$\widehat{e_0} \otimes \widehat{e_1} = \{ x \in \mathcal{F}_0 \otimes \mathcal{F}_1 \mid (e_0, e_1) \in x \},$$

where $\widehat{e_0} = \{ x_0 \in \mathcal{F}_0 \mid e_0 \in x_0 \}$ and $\widehat{e_1} = \{ x_1 \in \mathcal{F}_1 \mid e_1 \in x_1 \}$. It is clear that by using finite union and finite intersection we can get all compact stable neighborhood of $\mathcal{F}_0 \otimes \mathcal{F}_1$ out of compact neighborhoods of the form $A \otimes B$. It is also clear that for x, $y \in \mathcal{F}_0 \otimes \mathcal{F}_1$, $x \subseteq y$ if and only if $x \in (A \otimes B)$ implies $y \in (A \otimes B)$ for all $A \in \mathbf{SN}(\mathcal{F}_0)$, $B \in \mathbf{SN}(\mathcal{F}_1)$. Therefore, the construction given in Theorem 8.5 does meet the requirements set at the beginning of Section 8.2.

Proposition 8.8 *Suppose* A_1, $A_2 \in \mathbf{KSN}(\mathcal{F}_0)$, B_1, $B_2 \in \mathbf{KSN}(\mathcal{F}_1)$, *where* $\mathcal{F}_0$, $\mathcal{F}_1$ *are coherent spaces. Then*

$$\begin{aligned}&(A_1 \cap A_2) \otimes (B_1 \cap B_2)\\ &= (A_1 \otimes B_1) \cap (A_2 \otimes B_1) \cap (A_1 \otimes B_2) \cap (A_2 \otimes B_2).\end{aligned}$$

The proof goes through without using the compactness assumption.

Proof $\subseteq$: Suppose $x \in (A_1 \cap A_2) \otimes (B_1 \cap B_2)$. Then there exist $y_0 \in A_1 \cap A_2$, $y_1 \in B_1 \cap B_2$ such that $y_0 \times y_1 \Subset x$. It is obvious

$$x \in (A_1 \otimes B_1) \cap (A_2 \otimes B_1) \cap (A_1 \otimes B_2) \cap (A_2 \otimes B_2).$$

$\supseteq$: Assume

$$x \in (A_1 \otimes B_1) \cap (A_2 \otimes B_1) \cap (A_1 \otimes B_2) \cap (A_2 \otimes B_2).$$

By definition there are y_{11}, y_{21}, y_{12}, $y_{22} \in \mathcal{F}_0$ and z_{11}, z_{21}, z_{12}, $z_{22} \in \mathcal{F}_0$ such that $y_{ij} \in A_i$, $z_{ij} \in B_j$, and $y_{ij} \times z_{ij} \Subset x$ for $i = 1, 2$, $j = 1, 2$. We have, as A_1, B_1, A_2, B_2 are stable neighborhoods,

$$\begin{aligned}y_{11} \cap y_{12} \in A_1, \quad y_{21} \cap y_{22} \in A_2,\\ z_{11} \cap z_{21} \in B_1, \quad z_{12} \cap z_{22} \in B_2.\end{aligned}$$

Clearly

$$(y_{11} \cap y_{12}) \cup (y_{21} \cap y_{22}) \in A_1 \cap A_2,$$

$$(z_{11} \cap z_{21}) \cup (z_{12} \cap z_{22}) \in B_1 \cap B_2,$$

and,

$$\begin{aligned}&[(y_{11} \cap y_{12}) \cup (y_{21} \cap y_{22})] \times [(z_{11} \cap z_{21}) \cup (z_{12} \cap z_{22})]\\&\subseteq (y_{11} \times z_{11}) \cup (y_{12} \times z_{12}) \cup (y_{21} \times z_{21}) \cup (y_{22} \times z_{22})\\&\Subset x.\end{aligned}$$

Therefore $x \in (A_1 \cap A_2) \otimes (B_1 \cap B_2)$.

□

With respect to linear function space, we have, similar to Theorem 8.3,

Theorem 8.6 *Let $\mathcal{F}_0$, $\mathcal{F}_1$ be coherent spaces and*

$$A \in \mathbf{KSN}(\mathcal{F}_0), \ B \in \mathbf{KSN}(\mathcal{F}_1).$$

Then $A \multimap B \in \mathbf{KSN}(\mathcal{F}_0 \multimap \mathcal{F}_1)$, where

$$A \multimap B = \{ x \in \mathcal{F}_0 \multimap \mathcal{F}_1 \mid A \sqsubseteq_\mu (Pt\, x)^{-1} B \}.$$

Here $Pt\, x$ is a stable function determined by x, whose definition can be found in the proof of Proposition 6.3.

We remark that

$$(\widehat{e_0} \multimap \widehat{e_1}) = \{ x \in \mathcal{F}_0 \multimap \mathcal{F}_1 \mid (e_0, e_1) \in x \},$$

where $\widehat{e_0} = \{ x_0 \in \mathcal{F}_0 \mid e_0 \in x_0 \}$ and $\widehat{e_1} = \{ x_1 \in \mathcal{F}_1 \mid e_1 \in x_1 \}$. In fact, let $x \in \mathcal{F}_0 \multimap \mathcal{F}_1$ and $(e_0, e_1) \in x$. Then $\{ e_1 \} \subseteq Pt\, x \,\{ e_0 \}$, and $\{ e_0 \} \in \mu(Pt\, x)^{-1}\widehat{e_1}$. So $x \in (\widehat{e_0} \multimap \widehat{e_1})$. Suppose, on the other hand, that $x \in (\widehat{e_0} \multimap \widehat{e_1})$. We have $\{ e_0 \} \in \mu(Pt\, x)^{-1}\widehat{e_1}$. Therefore, $Pt\, x \,\{ e_0 \} \supseteq \{ e_1 \}$ and $(e_0, e_1) \in x$.

By this observation, each compact stable neighborhood of $\mathcal{F}_0 \multimap \mathcal{F}_1$ can be constructed out of $A \multimap B$ by using finite union and finite intersection. Also, for $x, y \in \mathcal{F}_0 \multimap \mathcal{F}_1$, $x \subseteq y$ if and only if

$$x \in (A \multimap B) \Longrightarrow y \in (A \multimap B)$$

for all $A \in \mathbf{KSN}(\mathcal{F}_0)$, $B \in \mathbf{KSN}(\mathcal{F}_1)$.

Thus the construction given in Theorem 8.6 is well-behaved.

Similar to Proposition 8.6, there is

Proposition 8.9 *Let A, B, C, D be stable neighborhoods and a a finite configuration. We have*

$$(A \cap B = \emptyset) \Longrightarrow (A \cup B) \multimap C = (A \multimap C) \cap (B \multimap C),$$
$$(A \cap B = \emptyset) \Longrightarrow \uparrow a \multimap (A \cup B) = (\uparrow a \multimap A) \cup (\uparrow a \multimap B),$$
$$(A \multimap C) \cap (B \multimap D) \subseteq (A \cap B) \multimap (C \cap D).$$

Note the rule $A' \subseteq A \;\&\; B \subseteq B' \Longrightarrow (A \multimap B) \subseteq (A' \multimap B')$ is not valid.

Proposition 8.10 *Let $x \in \mathcal{F}_0$ and $y \in \mathcal{F}_1$ be finite configurations and $\mathcal{F}_0$, $\mathcal{F}_1$ coherent spaces. We have*

$$e' \in y \Longrightarrow (\uparrow x \multimap \;\uparrow y) \subseteq \bigcup_{e \in x} (\widehat{e} \multimap \widehat{e'}).$$

Proof Suppose $w \in (\uparrow x \multimap \;\uparrow y)$ and $e' \in y$. Then $x \in \mu(Pt\, w)^{-1} \uparrow y$. We have $(Pt\, w)\,(x) \supseteq y$. This implies $e' \in (Pt\, w)\,(x)$. Hence

$$\exists e \in x.\, (e,\, e'\,) \in w,$$

which implies $w \in (\widehat{e} \multimap \widehat{e'})$, by the remark given after Theorem 8.6.

□

There is also a well-behaved construction with respect to exponential.

Theorem 8.7 *Let $\mathcal{F}$ be a coherent space and $A \in \mathbf{KSN}(\mathcal{F})$. Then*

$$!A \in \mathbf{KSN}(\,!\mathcal{F}),$$

where $!A = \{\, x \in !\mathcal{F} \mid x \cap \mu A \neq \emptyset \,\}$.

Proof Clearly $!A$ is upwards closed. It is open because the elements of μA are finite configurations. Now suppose x, $y \in !A$ and $x \uparrow y$. We have $x \cap y \in !\mathcal{F}$, $x \cap \mu A \neq \emptyset$, and $y \cap \mu A \neq \emptyset$. But $x \cup y \in !\mathcal{F}$, hence $(x \cap y) \cap \mu A \neq \emptyset$, because the elements of μA are pairwise inconsistent. So $x \cap y \in !A$ and $!A$ is a stable neighborhood.

To see $!A \in \mathbf{KSN}(!\mathcal{F})$ note that

$$!A = \bigcup_{a \in \mu A} !(\uparrow \{\, a \,\})$$

and each $!(\uparrow \{\, a \,\})$ is compact.

□

Notice that for x, $y \in !\mathcal{F}$, $x \subseteq y$ if and only if $x \in !A$ implies $y \in !A$ for all $A \in \mathbf{SN}(\mathcal{F})$.

Proposition 8.11 *Let* A, $B \in \mathbf{SN}(\mathcal{F})$ *and* $A \cap B = \emptyset$. *Then*

$$!(A \cup B) = (!A) \cup (!B).$$

Proof When $A \cap B = \emptyset$ we have $\mu(A \cup B) = \mu A \cup \mu B$.

□

Of course Definition 8.5 specializes down to coherent spaces. For coherent spaces $\mathcal{F}_0$ and $\mathcal{F}_1$ with $A \in \mathbf{KSN}(\mathcal{F}_0)$, $B \in \mathbf{KSN}(\mathcal{F}_1)$, we define

$$A \to B = \{\, x \in \mathcal{F}_0 \to \mathcal{F}_1 \mid A \sqsubseteq_\mu (Pt\, x)^{-1}B \,\}.$$

As a corollary of Theorem 8.3, we have $A \to B \in \mathbf{KSN}(\mathcal{F}_0 \to \mathcal{F}_1)$. It is well-known that there is an isomorphism $\ell : [\mathcal{F}_0 \to_s \mathcal{F}_1] \to [!\mathcal{F}_0 \multimap \mathcal{F}_1]$. Is it also true, under the isomorphism, that $A \to B \simeq (!A) \multimap B$, as one may expect? Of course. We have

Proposition 8.12 *Suppose*

$$\ell : [\mathcal{F}_0 \to_s \mathcal{F}_1] \to [!\mathcal{F}_0 \multimap \mathcal{F}_1]$$

is the isomorphism. Then

$$x \in A \to B \text{ if and only if } \ell x \in (!A) \multimap B,$$

where $A \in \mathbf{KSN}(\mathcal{F}_0)$, $B \in \mathbf{KSN}(\mathcal{F}_1)$.

Proof We have

$$\begin{aligned} x \in (A \to B) &\iff \mu A \subseteq \mu(Pt\, x)^{-1}B \\ &\iff \{\, \{a\} \mid a \in \mu A \,\} \subseteq \mu(Pt\, \ell x)^{-1}B \\ &\iff \mu(!A) \subseteq \mu(Pt\, \ell x)^{-1}B \\ &\iff (\ell x) \in (!A) \multimap B, \end{aligned}$$

where $\ell x = \{\, (\{u\}, e) \mid (u, e) \in x \,\}$. Note we used the fact that

$$\mu(!A) = \{\, \{a\} \mid a \in \mu A \,\}.$$

□

8.4 Parallel Products

This section studies the stable-neighborhood constructions in the categories $\mathbf{SEV}^*_{syn}$ and $\mathbf{SEV}_{syn}$. First we give some general results on how events and the relationships among them determine stable neighborhoods. We then focus on the partially synchronous product and synchronous product.

Theorem 8.8 *Let* $\underline{E} = (E, Con, \vdash)$ *be a stable event structure and let*

$$X \in Con.$$

Define

$$\widehat{X} = \{ x \in \mathcal{F}(\underline{E}) \mid X \subseteq x \}.$$

Then $\widehat{X}$ *is a stable neighborhood of* $(\mathcal{F}(\underline{E}), \subseteq)$.

Proof First we show that $\widehat{X}$ is Scott open. Obviously $\widehat{X}$ is upwards closed. Let

$$x_0 \subseteq x_1 \subseteq \cdots \subseteq x_i \cdots$$

be an increasing chain in $\mathcal{F}(\underline{E})$ such that $\bigcup_{i \in \omega} x_i \in \widehat{X}$. As X is finite, there must be some i such that $X \subseteq x_i$, or $x_i \in \widehat{X}$. Hence $\widehat{X}$ is open.

Let $x \uparrow y$ and $x, y \in \widehat{X}$. We have $x \cap y \in \mathcal{F}(\underline{E})$ as $\underline{E}$ is stable. Also $X \subseteq x \cap y$ as $X \subseteq x$ and $X \subseteq y$. Therefore $\widehat{X}$ is a stable neighborhood.

□

When $X = \{ e \}$ we write $\widehat{e}$ for $\widehat{\{ e \}}$. In particular $\widehat{e}$ is a stable neighborhood. But it is not necessarily compact. From Chapter 6 we know that the complete primes of $(\mathcal{F}(\underline{E}), \subseteq)$ for a stable event structure $\underline{E}$ are of the form $\lceil e \rceil_x$ where $e \in x$ and

$$\lceil e \rceil_x = \bigcap \{ y \in \mathcal{F}(\underline{E}) \mid e \in y \ \&\ y \subseteq x \}.$$

It is then not difficult to see that the minimal points of $\widehat{e}$ are all complete primes.

Now we consider some general constructions on stable neighborhoods.

Definition 8.6 *Suppose* $A, B \in \mathbf{SN}(D)$, *where* D *is a dI-domain. Define*

$$A \ll B = \{ x \in D \mid x \in B \ \&\ [\forall y \sqsubseteq x . y \in B \Rightarrow y \in A] \}.$$

$A \ll B$ reads 'B needs A' or 'A proceeds B'. A computation x has property $A \ll B$ if it has property B and at any earlier stage of the computation if property B is satisfied then property A is also satisfied.

Example 8.3 *Consider $(\mathcal{F}(\underline{E}), \subseteq)$, the dI-domain associated with a stable event structure $\underline{E}$, and let $A = \widehat{e}$, $B = \widehat{e'}$, where e, $e' \in E$. In this case*

$$x \in (\widehat{e} \ll \widehat{e'})$$

if and only if $e' \in x$ and for any $y \subseteq x$, $e' \in y$ implies $e \in x$. Comparing this with Definition 6.8 we find that $x \in (\widehat{e} \ll \widehat{e'})$ if and only if $e \leq_x e'$. Note according to Definition 6.8, $e \leq_x e'$ is only defined for those e, e' which belong to x.

Theorem 8.9 *Suppose D is a dI-domain and A, $B \in \mathbf{SN}(D)$. Then*

$$A \ll B \in \mathbf{SN}(D)$$

and, moreover,

$$(A \ll B) = \uparrow(A \cap \mu B).$$

Proof First we show that $(A \ll B)$ is upwards closed. Suppose

$$x \in (A \ll B), \ x \sqsubseteq z.$$

$x \in (A \ll B)$ implies $x \in B$. Hence $z \in B$. Let $y \sqsubseteq z$ and $y \in B$. We want to show that $y \in A$. x and y are compatible because $x \sqsubseteq z$ and $y \sqsubseteq z$. Hence $x \sqcap y \in B$. Since $x \in (A \ll B)$, $x \sqcap y \subseteq x$, and $x \sqcap y \in B$, we have $x \sqcap y \in A$. Therefore $y \in A$. Namely, $z \in (A \ll B)$.

Next we show that $(A \ll B) = \uparrow(A \cap \mu B)$, which implies $(A \ll B)$ is a stable neighborhood since it is then clear that $\mu(A \ll B) \subseteq \mu B$.

Let $x \in (A \ll B)$. Then $x \in B$. For some $c \in \mu B$, $c \sqsubseteq x$. Clearly $c \in A$, therefore $c \in (A \cap \mu B)$ and $x \in \uparrow(A \cap \mu B)$. Thus

$$(A \ll B) \subseteq \uparrow(A \cap \mu B).$$

On the other hand, if $x \in \uparrow(A \cap \mu B)$ then $x \sqsupseteq c$ for some $c \in (A \cap \mu B)$. We have $c \in (A \cap B)$ and for any $y \sqsubseteq c$, $y \in B$ implies $y = c$ since c is in μB, and c is in A. Hence $c \in (A \ll B)$. But $(A \ll B)$ is upwards closed, so $x \in (A \ll B)$. We have shown that

$$(A \ll B) \supseteq \uparrow(A \cap \mu B).$$

□

Corollary 8.1 *Let* $\underline{E} = (E, Con, \vdash)$ *be a stable event structure. Define, for two events* e, e',

$$(e \leq e') = \{ x \mid x \in \mathcal{F}(\underline{E}) \;\&\; e \leq_x e' \}.$$

$(e \leq e')$ *is a stable neighborhood of* $\mathcal{F}(\underline{E})$ *which is equal to* $\widehat{e} \ll \widehat{e'}$.

Abbreviate $(A \ll B) \cap (B \ll A)$ as $A \frown B$. Then a computation x has property $(A \frown B)$ if x has both properties A and B, and at any earlier stage of the computation, property A and property B either hold or not hold, at the *same* time. In other words, a computation x has property $(A \frown B)$ if property A and B start holding *simultaneously*. We can read $(A \frown B)$ as '*A synchronous-and B*'.

Definition 8.7 *Suppose* A, $B \in \mathbf{SN}(D)$, *where* D *is a dI-domain. Define*

$$A \not\ll B = \{ x \in D \mid x \in B \;\&\; [\exists y \sqsubseteq x. (y \in B \;\&\; y \notin A)] \}.$$

$(A \not\ll B)$ reads 'B independent of A'. Abbreviate $(A \not\ll B) \cap (B \not\ll A)$ as $A \smile B$. Then a computation x has property $(A \smile B)$ if x has both properties A and B, and at some earlier stage in the computation, property B holds but property A does not, and there is some earlier stage at which property A holds but property B does not. In other words, a computation x has property $(A \smile B)$ if at x both property A and B hold, but they started holding *independently*. We can read $(A \smile B)$ as '*A asynchronous-and B*'.

Example 8.4 *Consider* $(\mathcal{F}(\underline{E}), \subseteq)$, *the dI-domain associated with a stable event structure* $\underline{E}$, *and let* $A = \widehat{e}$, $B = \widehat{e'}$, *where* $e, e' \in E$. *We have* $x \in (\widehat{e} \smile \widehat{e'})$ *if and only if* $e' \in x$, $e \in x$ *and, for some* $y \subseteq x$, $e' \in y$ *but* $e \notin y$, *for some* $y' \subseteq x$, $e \in y'$ *but* $e' \notin y'$. *Hence* $x \in (\widehat{e} \smile \widehat{e'})$ *is a similar notion to* $e\, co_x\, e'$ *which people use in the literature to describe that events* e *and* e' *can occur concurrently in* x.

Theorem 8.10 *Suppose* D *is a dI-domain and* A, $B \in \mathbf{SN}(D)$. *Then* $A \not\ll B \in \mathbf{SN}(D)$ *and, moreover,*

$$(A \not\ll B) = \uparrow(\mu B \setminus A).$$

Proof It is enough to show that $(A \not\ll B) = \uparrow(\mu B \setminus A)$.

Let $x \in (A \not\leftarrow B)$. Then $x \in B$ and for some $y \sqsubseteq x$, $y \in B$ but $y \notin A$. For some $c \in \mu B$, $c \sqsubseteq y$. Clearly $c \notin A$. Therefore $c \in (\mu B \setminus A)$ and $x \in\uparrow(\mu B \setminus A)$. Thus

$$(A \not\leftarrow B) \subseteq\uparrow(\mu B \setminus A).$$

On the other hand, if $x \in\uparrow(\mu B \setminus A)$ then $x \sqsupseteq c$ for some $c \in (\mu B \setminus A)$. We have $x \in B$ and $c \sqsubseteq x$, $c \in B$ but $c \notin A$. So $x \in (A \not\leftarrow B)$. We have shown that

$$(A \not\leftarrow B) \supseteq\uparrow(\mu B \setminus A).$$

□

Let $A \in \mathbf{SN}(\mathcal{F}(\underline{E}_1))$ and $B \in \mathbf{SN}(\mathcal{F}(\underline{E}_2))$, where $\underline{E}_1$ and $\underline{E}_2$ are stable event structures. How can we construct, out of A and B, a reasonable stable neighborhood for the partially synchronous product $\underline{E}_1 \times \underline{E}_2$ given in Definition 6.17?

Definition 8.8 *Suppose $A \in \mathbf{SN}(\mathcal{F}(\underline{E}_1))$ and $B \in \mathbf{SN}(\mathcal{F}(\underline{E}_2))$, where $\underline{E}_1$ and $\underline{E}_2$ are stable event structures. Define*

$$\begin{aligned}(A \leftrightarrow_{ps} B) &= \{x \in \mathcal{F}(\underline{E}_1 \times \underline{E}_2) \mid \pi_1 x \in A \;\&\; \pi_2 x \in B \;\&\\ &\quad \forall y\, (y \in \mathcal{F}(\underline{E}_1 \times \underline{E}_2) \;\&\; y \subseteq x).\ \pi_1 y \in A \Leftrightarrow \pi_2 y \in B\,\}\end{aligned}$$

and

$$\begin{aligned}(A \leftrightarrow_{s} B) &= \{x \in \mathcal{F}(\underline{E}_1 \oplus \underline{E}_2) \mid \pi_1 x \in A \;\&\; \pi_2 x \in B \;\&\\ &\quad \forall y\, (y \in \mathcal{F}(\underline{E}_1 \oplus \underline{E}_2) \;\&\; y \subseteq x).\ \pi_1 y \in A \Leftrightarrow \pi_2 y \in B\,\}\end{aligned}$$

Theorem 8.11 *For stable event structures $\underline{E}_1$ and $\underline{E}_2$, if $A \in \mathbf{SN}(\mathcal{F}(\underline{E}_1))$ and $B \in \mathbf{SN}(\mathcal{F}(\underline{E}_2))$, then $(A \leftrightarrow_{ps} B) \in \mathbf{SN}(\mathcal{F}(\underline{E}_1 \times \underline{E}_2))$. Moreover, if $e_1 \in E_1$ and $e_2 \in E_2$, then $\widehat{e_1} \leftrightarrow_{ps} \widehat{e_2} = \widehat{(e_1, e_2)}$. The same result holds for the synchronous product, that is, if $A \in \mathbf{SN}(\mathcal{F}(\underline{E}_1))$ and $B \in \mathbf{SN}(\mathcal{F}(\underline{E}_2))$, then $(A \leftrightarrow_{s} B) \in \mathbf{SN}(\mathcal{F}(\underline{E}_1 \oplus \underline{E}_2))$. Moreover, if $e_1 \in E_1$ and $e_2 \in E_2$, then $\widehat{e_1} \leftrightarrow_{s} \widehat{e_2} = \widehat{(e_1, e_2)}$.*

Proof We check the case for the partially synchronous product. The proof for the synchronous product is similar.

Assume $x \in (A \leftrightarrow_{ps} B)$ and $x \subseteq z$, where $z \in \mathcal{F}(\underline{E}_1 \times \underline{E}_2)$. We have $\pi_1 z \in A$ and $\pi_2 z \in B$, clearly. For any $y \subseteq z$,

$$\begin{aligned}\pi_1 y \in A &\Longleftrightarrow^{*} \pi_1(x \cap y) \in A\\ &\Longleftrightarrow \pi_2(x \cap y) \in B \qquad (x \cap y \subseteq x \in (A \leftrightarrow_{ps} B))\\ &\Longleftrightarrow \pi_2 y \in B.\end{aligned}$$

($*$: we have, in this special case, $\pi_1(x \cap y) = (\pi_1 x) \cap (\pi_1 y)$ when $x \uparrow y$. This is because $\pi_1 e = \pi_1 e' \in (\pi_1 x) \cap (\pi_1 y)$ implies $e = e'$ with e, e' consistent.) Hence $z \in (A \leftrightarrow_{ps} B)$, which means $(A \leftrightarrow_{ps} B)$ is upwards closed.

Let

$$x_0 \subseteq x_1 \subseteq \cdots \subseteq x_n \subseteq \cdots$$

be a chain in $\mathcal{F}(\underline{E}_1 \times \underline{E}_2)$ such that $\bigcup_{i \in \omega} x_i \in (A \leftrightarrow_{ps} B)$. Obviously there is some x_i for which $\pi_1 x_i \in A$ and $\pi_2 x_i \in B$. It is then easy to see that $x_i \in (A \leftrightarrow_{ps} B)$. So $(A \leftrightarrow_{ps} B)$ is open. From x, $y \in (A \leftrightarrow_{ps} B)$ and $x \uparrow y$ we can easily deduce that $x \cap y \in (A \leftrightarrow_{ps} B)$.

To prove $(\widehat{e}_1 \leftrightarrow_{ps} \widehat{e}_2) = \widehat{(e_1,\, e_2)}$ it is clearly enough to show that

$$(\widehat{e}_1 \leftrightarrow_{ps} \widehat{e}_2) \subseteq \widehat{(e_1,\, e_2)}$$

because the inclusion in the other direction is obvious. Let

$$x \in (\widehat{e}_1 \leftrightarrow_{ps} \widehat{e}_2).$$

We have $e_1 \in \pi_1 x$ and $e_2 \in \pi_2 x$. Therefore there exist e, $e' \in x$, such that $\pi_1 e = e_1$ and $\pi_2 e' = e_2$. From the coincidence-freeness we deduce that $e = e' = (e_1,\, e_2)$.

□

At this point it is appropriate to ask whether all the criteria set at the beginning of Section 8.2 are satisfied for the constructions introduced in Definition 8.8. From Theorem 8.11 we know that for the synchronous product all the three conditions are satisfied. It is easy to see that the first and the third conditions hold. The second condition also holds since Theorem 8.11 concludes that it is possible to get the stable neighborhoods $\widehat{(e_1,\, e_2)}$. Clearly any prime stable neighborhood in $\mathbf{SN}(\mathcal{F}(\underline{E}_1 \oplus \underline{E}_2))$ is a finite intersection of stable neighborhoods of the form $\widehat{(e_1,\, e_2)}$.

For the partially synchronous product the first and the third conditions hold. The second condition, however, does not hold. This is because we cannot get stable neighborhoods of the form $\widehat{(e_1,\, *)}$ and $\widehat{(*,\, e_2)}$ in general. Based on Definition 8.7 we can introduce a similar notion as that is given in Definition 8.8, to capture the independence of events. Suppose $A \in \mathbf{SN}(\mathcal{F}(\underline{E}_1))$ and $B \in \mathbf{SN}(\mathcal{F}(\underline{E}_2))$, where $\underline{E}_1$ and $\underline{E}_2$ are stable event structures. Define

$$\begin{aligned}(A \not\lessdot_{ps} B) &= \{x \in \mathcal{F}(\underline{E}_1 \times \underline{E}_2) \mid \pi_2 x \in B \;\& \\ &\exists y\,(y \in \mathcal{F}(\underline{E}_1 \times \underline{E}_2) \;\&\; y \subseteq x).\ \pi_2 y \in B \;\&\; \pi_1 y \notin A\,\}\end{aligned}$$

and

$$\begin{aligned}(A \not\gtrdot_{ps} B) &= \{x \in \mathcal{F}(\underline{E}_1 \times \underline{E}_2) \mid \pi_1 x \in A \;\& \\ &\exists y\,(y \in \mathcal{F}(\underline{E}_1 \times \underline{E}_2) \;\&\; y \subseteq x).\ \pi_1 y \in A \;\&\; \pi_2 y \notin B\,\}.\end{aligned}$$

A stable neighborhood $\widehat{(e_1, *)}$ consists of all the configurations x for which $(e_1, *) \in x$. This means e_1 has occurred in x but it is independent of any event in E_2. By using the construction just introduced we can only have $\widehat{(e_1, *)}$ as

$$\widehat{(e_1, *)} = \bigcap_{e_2 \in E_2} (\widehat{e}_1 \not\gtrdot_{ps} \widehat{e}_2),$$

which is not necessarily a finite intersection. This suggests that we should use not only the finite disjunctions and finite conjunctions, but also limit form of quantifications like

$$\forall e \in E_2.\, (e_1 \text{ independent-of } e)$$

to express the fact that an event $(e_1, *)$ occurred in a configuration of the partially synchronous product.

Finally note that in both the partially synchronous product and synchronous product, a stable neighborhood $\widehat{(e_1, e_2)}$ is not necessarily compact as the following example shows.

Example 8.5 *Let* $\underline{E}_1 = (\{e, e'\},\ Con_0,\ \vdash_0)$, *where* $\{e, e'\} \in Con_0$ *and* $\{e\} \vdash_0 e'$, $\emptyset \vdash_0 e$; $\underline{E}_2 = (\mathcal{N},\ Con_1,\ \vdash_1)$, *where* $X \in Con_1$ *for* $X \Subset \mathcal{N}$ *and* $\emptyset \vdash_1 i$ *for every* $i \in \mathcal{N}$. *In the partially synchronous product or synchronous product,* $\widehat{(e', 0)}$ *is not compact because* $\{\,(e, i)\,\}$, $i \geq 1$, *are minimal points in* $\widehat{(e', 0)}$.

As remarked earlier, this means that there does not exist a finite prime normal form for parallel products. It does not imply, however, that there does not exist a decent logic on $\mathbf{SEV}^*_{syn}$ or $\mathbf{SEV}_{syn}$.

Chapter 9

Disjunctive Logics

This chapter studies the logic of stable domains. There are essentially two possible approaches for the logic of stable domains: one takes a disjunctive assertion language and the other takes a more conventional assertion language.

This chapter focuses more on the disjunctive approach. It is introduced in Section 9.1 through 9.4. Based on the material of Chapter 6 to Chapter 8, we introduce logical frameworks similar to that presented in Chapter 4 but for the category $\mathbf{COH}_l$ and the category **DI**. As before, type expressions, typed assertion languages, proof systems, and meta-predicates are the major components of the framework. However, assertions are generated in a different way here. The assertion language is equipped with a disjoint 'or', to cope with the disjunctive nature of stable neighborhoods. Also because of this disjoint nature, the assertions cannot be formulated by a simple grammar directly, as has been done conventionally; Instead we have to use an inductive definition combining a syntactic grammar with proof rules.

The second approach is discussed in Section 9.5. It adopts a conventional assertion language. However, to achieve completeness more proof rules have to be introduced. We illustrate a few of the new rules and explain how completeness can then be achieved.

Since there is no similar work to compare with, we could not say at the moment that the frameworks presented here definitely provides the best treatment of the logic of stable domains. For that reason a more appropriate title for this chapter would be 'Towards Disjunctive Logics'.

9.1 Disjunctive Assertions

The language of type expressions for coherent spaces is introduced as follows:

$$\sigma ::= O \mid \sigma \times \tau \mid \sigma \otimes \tau \mid \sigma + \tau \mid \sigma \multimap \tau \mid !\sigma \mid t \mid rec\, t.\sigma$$

where t is a type variable, and σ, τ ranges over type expressions.

Each closed type expression can be interpreted as a coherent space. The atomic type expression O is interpreted as the two point coherent space $\mathcal{O}$. To fully determine the interpretation it is enough to specify the interpretations for the type constructions. $\times$ is interpreted as the cartesian product, $\otimes$ the tensor product, $+$ the coproduct or sum, $\multimap$ the linear function space, ! the exponential, and $rec\, t.\sigma$ the recursively defined coherent spaces (see Section 7.5). Write $\mathcal{D}(\sigma)$ for the coherent space associated with a closed type expression σ specified in this way.

For each closed type expression we associate it with an assertion language. To characterize coherent spaces the assertions are interpreted as the compact stable neighborhoods of the space. It is intended that the assertions be disjunctive, in other words, when

$$\varphi \vee \psi$$

appear as an assertion it should be provable from the proof system that

$$\varphi \wedge \psi \leq \mathbf{f}.$$

There is actually a weaker notion than this, requiring that if

$$\varphi \vee \psi$$

is an assertion then

$$[\![\varphi \wedge \psi]\!] \subseteq \emptyset.$$

However since our proof system is complete, these two definitions coincide.

Surprisingly, such a simple requirement of disjunctiveness makes it impossible to specify the assertion language solely by a simple grammar. We introduce the assertion language by first introducing atomic assertions (tokens) and a related conflict relation, then giving syntactic rules and proof rules at the same time. This inductive definition starts with the atomic assertions. But when it is provable from the system that

$$\varphi \wedge \psi \leq \mathbf{f},$$

$\varphi \vee \psi$ becomes a proper assertion and we can prove more facts about these assertions, and can get even more assertions.

Notation. In this chapter all the index set such as I, J are finite.

Atomic assertions are specified by the following table. $\top$ is an atomic assertion of O which, by the semantic interpretation given later, stands for the top element of the Sierpinski space $\mathcal{O}$. As a special case (taking I to be the empty set) of the atomicity rule for the exponential, we have

$$\mathbf{A}_{!\sigma}(\ !\mathbf{t}_\sigma\ ),$$

which means $!\mathbf{t}_\sigma$ is an atomic assertion of type $!\sigma$. When the type is clear from the context, we just write $\mathbf{A}(\varphi)$. The next table specifies the inconsistency relation between atomic assertions. All assertions there are assumed to be atomic.

Atomic Assertions

$$\mathbf{A}_O(\top) \qquad \frac{\mathbf{A}_\sigma(\varphi)}{\mathbf{A}_{\sigma+\tau}(\mathit{inl}\,\varphi)} \qquad \frac{\mathbf{A}_\tau(\psi)}{\mathbf{A}_{\sigma+\tau}(\mathit{inr}\,\psi)}$$

$$\frac{\mathbf{A}_\sigma(\varphi)}{\mathbf{A}_{\sigma\times\tau}(\varphi\times\mathbf{t}_\tau)} \qquad \frac{\mathbf{A}_\tau(\psi)}{\mathbf{A}_{\sigma\times\tau}(\mathbf{t}_\sigma\times\psi)} \qquad \frac{\mathbf{A}_\sigma(\varphi) \quad \mathbf{A}_\tau(\psi)}{\mathbf{A}_{\sigma\otimes\tau}(\varphi\otimes\psi)}$$

$$\frac{\forall i, j \in I.\ \mathbf{A}_\sigma(\varphi_i)\ \&\ \neg(\varphi_i \# \varphi_j)}{\mathbf{A}_{!\sigma}(!\bigwedge_{i\in I}\varphi_i)} \qquad \frac{\mathbf{A}_\sigma(\varphi) \quad \mathbf{A}_\tau(\psi)}{\mathbf{A}_{\sigma\multimap\tau}(\varphi\multimap\psi)} \qquad \frac{\mathbf{A}_{\sigma[\mathit{rec}\,t.\sigma/t]}(\varphi)}{\mathbf{A}_{\mathit{rec}\,t.\sigma}(\varphi)}$$

Atomic Inconsistency

$$\mathit{inl}\,\varphi \# \mathit{inr}\,\psi \qquad \frac{\varphi \# \varphi'}{\mathit{inl}\,\varphi \# \mathit{inl}\,\varphi'} \qquad \frac{\psi \# \psi'}{\mathit{inr}\,\psi \# \mathit{inr}\,\psi'}$$

$$\frac{\varphi \# \varphi'}{\varphi\times\mathbf{t} \# \varphi'\times\mathbf{t}} \qquad \frac{\psi \# \psi'}{\mathbf{t}\times\psi \# \mathbf{t}\times\psi'} \qquad \frac{\exists i \in I\, \exists j \in J.\ \varphi_i \# \psi_j}{!\bigwedge_{i\in I}\varphi_i \# !\bigwedge_{j\in J}\psi_j}$$

$$\frac{\varphi \# \varphi'}{\varphi\otimes\psi \# \varphi'\otimes\psi'} \qquad \frac{\psi \# \psi'}{\varphi\otimes\psi \# \varphi'\otimes\psi'}$$

$$\frac{\neg(\varphi \# \varphi') \quad \psi \# \psi'}{\varphi\multimap\psi \# \varphi'\multimap\psi'} \qquad \frac{\neg(\varphi \# \varphi') \quad \neg(\varphi \sim \varphi') \quad \psi \sim \psi'}{\varphi\multimap\psi \# \varphi'\multimap\psi'}$$

On atomic assertions the *similarity relation* $\sim$ almost means the syntactic equality ($\equiv$). The subtlety comes from the atomic assertions of type $!\sigma$. We do not want to distinguish an assertion $!(\varphi_0 \wedge \varphi_1)$ with the assertion $!(\varphi_1 \wedge \varphi_0)$ in the similarity relation. Hence in an atomic assertion $!\bigwedge_{i\in I}\varphi_i$ of type $!\sigma$, the order of the conjunctives is ignored by $\sim$. This has the same effect as treating $\sim$ as the syntactic equality except that every assertion $!\bigwedge_{i\in I}\varphi_i$ of type $!\sigma$ is understood as $!(\{\varphi_i \mid i \in I\})$. It is intended that two atomic assertions are similar if and only if they are logically equivalent.

Note in the foregoing tables not only $\#$ and $\sim$ are used, but also their negations. That should not cause any problem because they are extremely simple relations. Once $\#$ and $\sim$ are given, their negations should also be considered as given.

Based on atomic assertions and the inconsistency relation on them it is now possible to specify the syntax of disjunctive assertions in general.

Assertions

$$\mathbf{t}, \mathbf{f} : \sigma \qquad\qquad \frac{\mathbf{A}_\sigma(\varphi)}{\varphi : \sigma}$$

$$\frac{\varphi,\ \psi : \sigma}{\varphi \wedge \psi : \sigma} \qquad\qquad \frac{\varphi \wedge \psi \leq \mathbf{f} \qquad \varphi,\ \psi : \sigma}{\varphi \vee \psi : \sigma}$$

$$\frac{\varphi : \sigma \qquad \psi : \tau}{\varphi \times \psi : \sigma \times \tau} \qquad\qquad \frac{\varphi : \sigma \qquad \psi : \tau}{\varphi \otimes \psi : \sigma \otimes \tau}$$

$$\frac{\varphi : \sigma}{\mathit{inl}\,\varphi : \sigma + \tau} \qquad\qquad \frac{\psi : \tau}{\mathit{inr}\,\psi : \sigma + \tau}$$

$$\frac{\varphi : \sigma}{!\varphi : !\sigma} \qquad\qquad \frac{\varphi : \sigma \qquad \psi : \tau}{\varphi \multimap \psi : \sigma \multimap \tau}$$

$$\frac{\varphi : \sigma[\mathit{rec}\, t.\sigma/t]}{\varphi : \mathit{rec}\, t.\sigma}$$

The rule that ensures disjunctiveness is

$$\frac{\varphi \wedge \psi \leq \mathbf{f} \qquad \varphi,\ \psi : \sigma}{\varphi \vee \psi : \sigma},$$

which says $\varphi \vee \psi$ is an assertion if we can derive the formula $\varphi \wedge \psi \leq \mathbf{f}$ The assertion language is the minimal set of assertions closed under these rules. On the other hand, if $\varphi \vee \psi$ is an assertion it must be the case that

$\vdash \varphi \wedge \psi \leq \mathbf{f}$. Write $\mathcal{B}_\sigma$ for the set of disjunctive assertions of type σ built up this way.

When $\varphi : \sigma$ we say φ is well-formed. For example $\mathbf{t} \vee \mathbf{f}$ is a well-formed assertion but $\mathbf{t} \vee \mathbf{t}$ is not. From now on we assume that all assertions are well-formed whenever they appear in the proof system.

Similar to the logic of SFP domains we also have a collection of propositional rules. Note however, (#) is new.

Propositional Rules

(t) $\quad \varphi \leq \mathbf{t}$

(f) $\quad \mathbf{f} \leq \varphi$

(Ref) $\quad \varphi \leq \varphi$

(#) $\quad \dfrac{\varphi \# \psi}{\varphi \wedge \psi \leq \mathbf{f}}$

(Trans) $\quad \dfrac{\varphi \leq \varphi' \quad \varphi' \leq \varphi''}{\varphi \leq \varphi''}$

$(\leq - =)$ $\quad \dfrac{\varphi \leq \psi \quad \psi \leq \varphi}{\varphi = \psi}$

$(= - \leq)$ $\quad \dfrac{\varphi = \varphi'}{\varphi \leq \varphi'} \qquad \dfrac{\varphi = \varphi'}{\varphi' \leq \varphi}$

$(\wedge - \leq)$ $\quad \varphi \wedge \varphi' \leq \varphi \qquad \varphi \wedge \varphi' \leq \varphi'$

$(\leq - \wedge)$ $\quad \dfrac{\varphi \leq \varphi' \quad \varphi \leq \varphi''}{\varphi \leq \varphi' \wedge \varphi''}$

$(\vee - \leq)$ $\quad \dfrac{\varphi \leq \varphi' \quad \psi \leq \varphi'}{\varphi \vee \psi \leq \varphi'}$

$(\leq - \vee)$ $\quad \varphi \leq \varphi' \vee \varphi \qquad \varphi' \leq \varphi' \vee \varphi$

$(\wedge - \vee)$ $\quad \varphi \wedge (\varphi_1 \vee \varphi_2) \leq (\varphi \wedge \varphi_1) \vee (\varphi \wedge \varphi_2)$

It is necessary to require that all the assertions appear in a rule be well-formed when we applying the rule. For example, with respect to the following axiom

$$\varphi \wedge (\varphi_1 \vee \varphi_2) \leq (\varphi \wedge \varphi_1) \vee (\varphi \wedge \varphi_2)$$

the fact that $(\varphi \wedge \varphi_1) \vee (\varphi \wedge \varphi_2)$ is well-formed does not imply $\varphi_1 \vee \varphi_2$ is also well formed: simply take the assertion to be $(\mathbf{f} \wedge \mathbf{t}) \vee (\mathbf{f} \wedge \mathbf{t})$. This is a well-formed assertion, but not $\mathbf{t} \vee \mathbf{t}$.

The proof systems associated with sum and product are similar to those given for **SFP**. So no explanation is needed for them.

Sum

$$(\mathit{inl}-\leq) \quad \frac{\varphi \leq \psi}{\mathit{inl}\,\varphi \leq \mathit{inl}\,\psi} \qquad (\mathit{inr}-\leq) \quad \frac{\varphi \leq \psi}{\mathit{inr}\,\varphi \leq \mathit{inr}\,\psi}$$

$$(\mathit{inl}-\wedge) \quad \mathit{inl}\,(\textstyle\bigwedge_{i\in I}\varphi_i) = \bigwedge_{i\in I}\mathit{inl}\,\varphi_i$$

$$(\mathit{inr}-\wedge) \quad \mathit{inr}\,(\textstyle\bigwedge_{i\in I}\psi_i) = \bigwedge_{i\in I}\mathit{inr}\,\psi_i$$

$$(\mathit{inl}-\vee) \quad \mathit{inl}\,(\textstyle\bigvee_{i\in I}\varphi_i) = \bigvee_{i\in I}\mathit{inl}\,\varphi_i$$

$$(\mathit{inr}-\vee) \quad \mathit{inr}\,(\textstyle\bigvee_{i\in I}\psi_i) = \bigvee_{i\in I}\mathit{inr}\,\psi_i$$

Note that from the rules for Atomic Inconsistency we have $\mathit{inl}\varphi \,\#\, \mathit{inr}\psi$, and by (#) of the Propositional Rules, we can derive (by the prime normal form theorem later) the rule

$$\frac{(\varphi)\downarrow \qquad (\psi)\downarrow}{\mathit{inl}\varphi \wedge \mathit{inr}\psi}$$

where $(\ )\downarrow$, a convengence predicate, will be introduced in just a moment. The proof systems for tensor product, product, exponential, and linear function space are introduced below.

Tensor Product

$$(\otimes-\mathbf{t}) \quad \frac{\mathbf{P}(\varphi)}{\varphi \otimes \mathbf{t}_\tau = \mathbf{t}_{\sigma\otimes\tau}} \qquad \frac{\mathbf{P}(\psi)}{\mathbf{t}_\sigma \otimes \psi = \mathbf{t}_{\sigma\otimes\tau}}$$

$$(\otimes-\leq) \quad \frac{\psi \leq \psi' \quad \varphi \leq \varphi'}{\psi \otimes \varphi \leq \psi' \otimes \varphi'}$$

$$(\otimes-\wedge) \quad \varphi \otimes (\psi_1 \wedge \psi_2) = \varphi \otimes \psi_1 \wedge \varphi \otimes \psi_2$$

$$(\varphi_1 \wedge \varphi_2) \otimes \psi = \varphi_1 \otimes \psi \wedge \varphi_2 \otimes \psi$$

$$(\otimes-\vee) \quad \varphi \otimes \textstyle\bigvee_{i\in I}\psi_i = \bigvee_{i\in I}\varphi \otimes \psi_i$$

$$(\textstyle\bigvee_{i\in I}\varphi_i) \otimes \psi = \bigvee_{i\in I}\varphi_i \otimes \psi$$

Here **P** is a predicate which captures prime assertions – the definition is given shortly. Note the special property of tensor product: for assertions with one of the component prime and the other being **t** the whole assertion is equivalent to **t**. This is expressed by the rules $(\otimes - \mathbf{t})$, which are sound because according to Theorem 8.5,

$$A \otimes B = \{ x \in \mathcal{F}_0 \otimes \mathcal{F}_1 \mid \exists x_0 \in A, x_1 \in B.\ x_0 \times x_1 \Subset x \}.$$

As a particular instance,

$$\begin{aligned}&\uparrow a \otimes \uparrow \perp \\ &= \{ x \in \mathcal{F}_0 \otimes \mathcal{F}_1 \mid \exists x_0 \supseteq a,\ x_1 \supseteq \emptyset.\ x_0 \times x_1 \Subset x \} \\ &= \{ x \in \mathcal{F}_0 \otimes \mathcal{F}_1 \mid \exists x_0 \supseteq a.\ x_0 \times \emptyset \Subset x \} \\ &= \mathcal{F}_0 \otimes \mathcal{F}_1.\end{aligned}$$

Note as a special case of axiom $(\otimes - \vee)$ we have $\varphi \otimes \mathbf{f} = \mathbf{f}$ and $\mathbf{f} \otimes \psi = \mathbf{f}$.

The rules for product are the same as before.

Product

$$(\times - \leq) \quad \frac{\psi \leq \psi' \quad \varphi \leq \varphi'}{\psi \times \varphi \leq \psi' \times \varphi'}$$

$$(\times - \vee) \quad \varphi \times (\textstyle\bigvee_{i \in I} \psi_i) = \bigvee_{i \in I} \varphi \times \psi_i$$

$$(\textstyle\bigvee_{i \in I} \varphi_i) \times \psi = \bigvee_{i \in I} \varphi_i \times \psi$$

$$(\times - \wedge) \quad \textstyle\bigwedge_{i \in I} \varphi_i \times \psi_i = (\bigwedge_{i \in I} \varphi_i) \times (\bigwedge_{i \in I} \psi_i)$$

There is only one axiom for exponential. As a special case of it we have $!\mathbf{f} = \mathbf{f}$. We do not have, however,

$$!(\varphi_0 \wedge \varphi_1) = !(\varphi_0) \wedge !(\varphi_1),$$

neither $!\mathbf{t} = \mathbf{t}$.

Exponential

$$(! - \vee) \qquad !(\textstyle\bigvee_{i \in I} \varphi_i) = \bigvee_{i \in I} !(\varphi_i)$$

$$(! - =) \qquad \frac{\varphi = \psi}{!\varphi = !\psi}$$

Note that we do not have

$$\frac{\varphi \leq \psi}{!\varphi \leq !\psi}.$$

We should have it for the logic of stable domains (see discussions after Theorem 6.13), however. This is because of the particular way in which the exponential works.

Linear Function Space

$(\multimap -\mathbf{f})$
$$\frac{\varphi\downarrow}{(\varphi\multimap\mathbf{t})\leq\mathbf{f}}$$

$(\multimap)$
$$\frac{\lceil\varphi\rceil\neq\lceil\varphi'\rceil \qquad \mathbf{P}(\varphi\wedge\varphi') \qquad \mathbf{P}(\psi)}{(\varphi\multimap\psi)\wedge(\varphi'\multimap\psi)\leq\mathbf{f}}$$

$(\multimap -\leq)$
$$\frac{a\in\lceil\psi\rceil \qquad \mathbf{P}(\varphi)}{\varphi\multimap\psi\leq\bigvee_{b\in\lceil\varphi\rceil}(b\multimap a)}$$

$(\multimap -\wedge)$
$$\bigwedge_{i\in I}(\varphi_i\multimap\psi_i)\leq(\bigwedge_{i\in I}\varphi_i)\multimap(\bigwedge_{i\in I}\psi_i)$$

$(\multimap -\vee)$
$$\frac{\mathbf{P}(\varphi)}{\varphi\multimap(\bigvee_{i\in I}\psi_i)=\bigvee_{i\in I}(\varphi\multimap\psi_i)}$$
$$(\bigvee_{i\in I}\varphi_i)\multimap\psi=\bigwedge_{i\in I}(\varphi_i\multimap\psi)$$

The rules for the linear function space need some explanation. We have the rule $(\multimap -\mathbf{f})$ because for any stable function $f : D\to E$,

$$\mu f^{-1}(E)=\bot_D.$$

Therefore if φ converges it is not possible for

$$\mu\,[\![\varphi]\!]\subseteq\mu f^{-1}([\![\mathbf{t}]\!])$$

to hold. The second rule, $(\multimap)$, reflects the construction given in Definition 6.21. The third rule is a close translation of Proposition 8.10. The rest of the rules are quite standard.

As special cases of the rules for linear function space, we have $\mathbf{f}\multimap\varphi=\mathbf{t}$, $\mathbf{t}=\mathbf{t}\multimap\mathbf{t}$, and $\varphi\multimap\mathbf{f}=\mathbf{f}$ for a prime assertion φ. Note that by $(\multimap -\leq)$, $\mathbf{t}\multimap\psi=\mathbf{f}$ if ψ converges. By $(\multimap -\mathbf{f})$, $\varphi\multimap\mathbf{t}=\mathbf{f}$ provided $\varphi\downarrow$.

Now the stable function space $\sigma\to_s\tau$ corresponds to the type $!\sigma\multimap\tau$. By Proposition 8.12, we can use $!\varphi\multimap\psi$ as assertions for the type $\sigma\to_s\tau$. Therefore our proof system can also treat stable function space.

Some of the rules were attached some side conditions. $\mathbf{P}(\varphi)$ reads φ is *prime*. $\varphi\downarrow$ reads φ is *convergent*. $\lceil\varphi\rceil$ is an operator called the *atomizer*, which returns a set of atomic assertions $\lceil\varphi\rceil$ not similar to one another, for a given prime assertion φ. Attaching side conditions to rules is not new;

we have seen similar side conditions used in predicate calculi and lambda calculi, like 'x does not occur free in M'. Those conditions are usually purely syntactic and easily checked.

Here are the formal definitions of these syntactic predicates.

$$\mathbf{P}(\varphi) \Longleftrightarrow^{\mathrm{def}} \varphi \equiv \bigwedge_{i\in I} \varphi_i \ \&\ \forall i \in I.\, \mathbf{A}(\varphi_i) \ \&\ \forall i, j \in I.\, \neg(\varphi_i \# \varphi_j),$$
$$\varphi\downarrow \Longleftrightarrow^{\mathrm{def}} \mathbf{P}(\varphi) \ \&\ \varphi \equiv \bigwedge_{i\in I} \varphi_i \ \&\ I \neq \emptyset.$$

Let φ be a prime assertion, *i.e.* $\mathbf{P}(\varphi)$. Then there are assertions φ_i, $i \in I$, such that $\varphi \equiv \bigwedge_{i\in I} \varphi_i$ and $\forall i \in I.\, \mathbf{A}(\varphi_i)$. We define $\lceil\varphi\rceil = \{\, \varphi_i \mid i \in I \,\}/\!\sim$. Here $\sim$ is the similarity relation introduced earlier.

We now give a semantics for the proof system, with assertions interpreted as stable neighborhoods, and entailment as set inclusion. For each closed type expression σ we define an *interpretation function*

$$[\![\ \ ]\!]_\sigma : \mathcal{B}_\sigma \to \mathbf{KSN}(\mathcal{D}(\sigma))$$

in the following structural way.

For each closed type expression σ, let

$$\begin{aligned}
&[\![\mathbf{t}]\!]_\sigma = \mathcal{D}(\sigma),\\
&[\![\mathbf{f}]\!]_\sigma = \emptyset,\\
&[\![\top]\!]_O = \top,\\
&[\![\varphi \vee \psi]\!]_\sigma = [\![\varphi]\!]_\sigma \cup [\![\psi]\!]_\sigma,\\
&[\![\varphi \wedge \psi]\!]_\sigma = [\![\varphi]\!]_\sigma \cap [\![\psi]\!]_\sigma.
\end{aligned}$$

With respect to type constructions we define

$$\begin{aligned}
&[\![\mathit{inl}\,\varphi]\!]_{\sigma+\tau} = \{(0,u) \mid u \in [\![\varphi]\!]_\sigma \setminus \{\emptyset\}\} \cup \{x \in \mathcal{D}(\sigma+\tau) \mid \emptyset \in [\![\varphi]\!]_\sigma\},\\
&[\![\mathit{inr}\,\varphi]\!]_{\sigma+\tau} = \{(1,u) \mid u \in [\![\varphi]\!]_\tau \setminus \{\emptyset\}\} \cup \{x \in \mathcal{D}(\sigma+\tau) \mid \emptyset \in [\![\varphi]\!]_\tau\},\\
&[\![\varphi \times \psi]\!]_{\sigma\times\tau} = \{(0,u) \mid u \in [\![\varphi]\!]_\sigma\} \cup \{(1,v) \mid v \in [\![\psi]\!]_\tau\},\\
&[\![\varphi \otimes \psi]\!]_{\sigma\otimes\tau} = [\![\varphi]\!]_\sigma \otimes [\![\psi]\!]_\tau,\\
&[\![!\varphi]\!]_{!\sigma} = ![\![\varphi]\!]_\sigma,\\
&[\![\varphi \multimap \psi]\!]_{\sigma\multimap\tau} = [\![\varphi]\!]_\sigma \multimap [\![\psi]\!]_\tau),\\
&[\![\varphi]\!]_{rect.\,\sigma} = \{\, \epsilon_\sigma(u) \mid u \in [\![\varphi]\!]_{\sigma[(rect.\,\sigma)\backslash t]}\},
\end{aligned}$$

where $\epsilon_\sigma : [\, \mathcal{D}(\sigma[(rect.\,\sigma)\backslash t]) \to \mathcal{D}(rect.\,\sigma)\,]$ is the isomorphism arising form the solution to the domain equation associated with type $rec\,t.\,\sigma$.

9.2 Completeness and Expressiveness

In this section we show that the proof system introduced for $\mathbf{COH}_l$ is sound and complete with respect to the interpretation given above, for *closed types*. For expressiveness we prove that any compact stable neighborhood is expressible in the logic. Some definitions first.

Definition 9.1 *Let σ be a closed type expression. For $\varphi, \psi \in \mathcal{B}_\sigma$, write $\models_\sigma \varphi \leq \psi$ if $[\![\varphi]\!]_\sigma \subseteq [\![\psi]\!]_\sigma$. We write $\vdash \varphi \leq_\sigma \psi$ if $\varphi \leq_\sigma \psi$ can be derived from the proof system given in the previous section. The proof system is called sound if $\vdash \varphi \leq_\sigma \psi$ implies $\models \varphi \leq_\sigma \psi$. It is* complete *if $\models \varphi \leq_\sigma \psi$ implies $\vdash \varphi \leq_\sigma \psi$. An axiom is valid if it is a valid formula. A rule is sound if it produces valid a formula from valid formulae.*

Proposition 9.1 *There is an isomorphism between atomic assertions of σ and tokens of $\mathcal{D}(\sigma)$ such that for any φ, ψ, atomic assertions of type σ, $\varphi \# \psi$ if and only if*

$$[\![\varphi]\!]_\sigma \cap [\![\psi]\!]_\sigma \subseteq \emptyset,$$

and $\varphi \sim \psi$ if and only if

$$[\![\varphi]\!]_\sigma = [\![\psi]\!]_\sigma.$$

Proof By a structural induction on the types.

□

The following proposition is routine to prove. It confirms that the interpretation of a prime assertion is a complete prime in the lattice of stable neighborhoods, the interpretation of an atomic assertion is a very prime neighborhood, one that is determined by a complete prime, and the interpretation of a convergent assertion does not contain the bottom.

Proposition 9.2

1. *Suppose $\varphi \in \mathcal{B}_\sigma$ and $\mathbf{P}(\varphi)$. Then there exists a finite element $x \in \mathcal{D}(\sigma)$ such that $[\![\varphi]\!]_\sigma = \uparrow x$.*
2. *Suppose φ is an atomic assertion of type σ. Then there exists $p \in \mathcal{D}(\sigma)$, a complete prime, such that $[\![\varphi]\!]_\sigma = \uparrow p$.*
3. *Suppose $\varphi \in \mathcal{B}_\sigma$ and $\varphi\downarrow$. Then $\mathcal{D}_\sigma \not\subseteq [\![\varphi]\!]_\sigma$.*

It is clear that a proof system is sound if and only if all its axioms are valid and rules sound. Theorem 9.1 establishes the soundness of the proof system.

Theorem 9.1

1. *The logical axioms are valid and logical rules sound.*
2. *The axioms for sum, product, tensor product, exponential, and hspace2cmlinear function space are valid.*
3. *The rules for sum, product, tensor product, exponential, and linear function space are sound.*

Proof We check the cases for tensor product, exponential, and linear function space.

The soundness of $(\otimes - \mathbf{t})$ follows from the definition given in Theorem 8.5: for two stable neighborhoods A and B, if $A \neq \emptyset$ and $\emptyset \in B$, then clearly $\emptyset \in A \otimes B$ as $x \times \emptyset = \emptyset$. The soundness of $(\otimes - \leq)$ and $(\otimes - \vee)$ is trivial while the soundness of $(\otimes - \wedge)$ follows from Proposition 8.8.

The soundness of $(! - =)$ for exponential follows from Proposition 8.11.

The soundness of $(\multimap - \vee)$, $(\multimap - \wedge)$ follows from Proposition 8.9 and Proposition 9.2. The soundness fo $(\multimap - \leq)$ follows from Proposition 8.10, Proposition 9.1 and Proposition 9.2, while the soundness of $(\multimap - \mathbf{f})$ follows from the definition given in Theorem 8.6 and Proposition 9.2. The rule $(\multimap)$ is sound since if f is a stable function and $x, y \in \mu f^{-1}(A)$ with $x \uparrow y$, then we must have $x = y$.

□

Remark. In proofs of this section we need not check the case for recursively defined types, because all the time we are dealing with finite sets of assertions, and these assertions can always be considered as of some finite type.

Definition 9.2 *Write $\mathcal{P}_\sigma$ for the proof system associated with type σ. $\mathcal{P}_\sigma$ is called prime complete if it has property p_0, prime normal if it has property p_1, and complete if it has property p_2, where*

$$(p_0) \qquad \forall \varphi, \psi : \sigma. (\mathbf{P}(\varphi) \,\&\, \mathbf{P}(\psi) \,\&\, [\![\varphi]\!]_\sigma \subseteq [\![\psi]\!]_\sigma \Longrightarrow \vdash \varphi \leq \psi),$$

$$(p_1) \qquad \varphi : \sigma \Longrightarrow \exists \{ \varphi_i \mid i \in I \}. \forall i \in I. \mathbf{P}(\varphi_i) \,\&\, \vdash \varphi = \bigvee_{i \in I} \varphi_i,$$

$$(p_2) \qquad \forall \varphi, \psi : \sigma. ([\![\varphi]\!] \subseteq [\![\psi]\!] \Longrightarrow \vdash \varphi \leq \psi).$$

In (p_1), $\bigvee_{i \in I} \varphi_i$ is called a prime normal form of φ.

Clearly $\mathcal{P}_O$ has property p_0, p_1, and p_2. The proof for the completeness of the system is achieved by showing that each type construction preserves property (p_0), (p_1), and (p_2), by a sequence of propositions. We omit the cases for sum and product, since they are not new. Note that for each case the proof of (p_2) is routine. It follows from (p_0) and (p_1) directly.

Proposition 9.3 *Tensor product preserves* p_0, p_1, *and* p_2.

Proof (p_0). Suppose $\varphi \otimes \psi$, $\varphi' \otimes \psi' : \sigma \times \tau$ are prime and

$$[\![\varphi \otimes \psi]\!] \subseteq [\![\varphi' \otimes \psi']\!].$$

This implies, by definition,

$$\varphi \otimes \psi \equiv \bigwedge_{i \in I} \varphi_i \otimes \psi_i$$

and

$$\varphi' \otimes \psi' \equiv \bigwedge_{j \in J} \varphi'_j \otimes \psi'_j,$$

where φ_i, ψ_i, φ'_j, and ψ'_j are atomic. By Proposition 9.1, for any $j \in J$, there is some $i \in I$,

$$[\![\varphi_i \otimes \psi_i]\!] = [\![\varphi'_j \otimes \psi'_j]\!].$$

Therefore $[\![\varphi_i]\!] = [\![\varphi'_j]\!]$ and $[\![\psi_i]\!] = [\![\psi'_j]\!]$. By assumption, $\vdash \varphi_i = \varphi'_j$ and $\vdash \psi_i = \psi'_j$. Hence $\vdash \varphi_i \otimes \psi_i = \varphi'_j \otimes \psi'_j$, by ($\otimes - \leq$). Using logical rules we get $\vdash \varphi \otimes \psi \leq \varphi' \otimes \psi'$.

(p_1) is routine.

□

Proposition 9.4 *Exponential preserves* p_0, p_1, *and* p_2.

Proof (p_0). Suppose

$$[\![\bigwedge_{i \in I} !\varphi_i]\!] \subseteq [\![\bigwedge_{j \in J} !\psi_j]\!],$$

where $!\varphi_i$'s and $!\psi_j$'s are atomic. For any $j \in J$, there is some $i \in I$, $[\![\varphi_i]\!] = [\![\psi_j]\!]$. By assumption, $\vdash \varphi_i = \psi_j$. Hence $\vdash !\varphi_i = !\psi_j$, by ($! - =$). Using logical rules we get $\vdash \bigwedge_{i \in I} !\varphi_i \leq \bigwedge_{j \in J} !\psi_j$.

(p_1) is easy.

□

Proposition 9.5 *Linear function space preserves p_0, p_1, and p_2.*

Proof The proof for p_1 is interesting in this case. Clearly, by $(\multimap - \vee)$, $(\multimap - \mathbf{f})$, and the logical rules one can reduce each assertion of a linear function space to a disjunctive form where each disjunct is a conjunction of assertions of the form $\varphi \multimap \psi$, with φ and ψ prime. It is enough to show that every such $\varphi \multimap \psi$ is reducible to a prime normal form. By $(\multimap - \leq)$ and logical rules we have

$$\begin{aligned} \vdash \varphi \multimap \psi \ &\leq \bigwedge_{a \in \lceil \psi \rceil} \bigvee_{b \in \lceil \varphi \rceil} (b \multimap a) \\ &\leq \bigvee_{\kappa : \lceil \psi \rceil \to \lceil \varphi \rceil} \ \bigwedge_{a \in \lceil \psi \rceil} (\kappa(a) \multimap a) \\ &\leq^{*} \bigvee_{\kappa \text{ onto}} \ \bigwedge_{a \in \lceil \psi \rceil} (\kappa(a) \multimap a) \\ &\leq^{**} \varphi \multimap \psi \end{aligned}$$

$*$: This is because for those κ's which are not onto, we can assume, say,

$$b_0 \in \lceil \varphi \rceil \setminus \kappa(\lceil \psi \rceil).$$

Therefore

$$\vdash (\varphi \multimap \psi) \wedge \bigwedge_{a \in \lceil \psi \rceil} (\kappa(a) \multimap a) \leq (\varphi \multimap \psi) \wedge (\varphi' \multimap \psi) \leq \mathbf{f}$$

by $(\multimap - \wedge)$ and $(\multimap)$, where $\varphi' \equiv \bigwedge_{a \in \lceil \psi \rceil} \kappa(a)$. Now use the fact that if $\varphi \wedge \psi_1 \leq \mathbf{f}$ and $\varphi \leq \psi_0 \vee \psi_1$, then

$$\varphi \leq \varphi \wedge (\psi_0 \vee \psi_1) \leq \varphi \wedge \psi_0.$$

Hence $\varphi \leq \psi_0$.

$**$: For those κ's which are onto, we clearly have, by $(\multimap - \wedge)$,

$$\vdash \bigwedge_{a \in \lceil \psi \rceil} (\kappa(a) \multimap a) \leq \varphi \multimap \psi.$$

Hence

$$\vdash \varphi \multimap \psi = \bigvee_{\kappa : \lceil \psi \rceil \to \lceil \varphi \rceil \text{ onto}} \ \bigwedge_{a \in \lceil \psi \rceil} (\kappa(a) \multimap a)$$

□

In summary we have proved

Theorem 9.2 *The proof system for the logic of coherent spaces is complete.*

The expressive result is what one can expect:

Theorem 9.3 *Let σ be a closed type expression. Then*

$$[\![\ ]\!]_\sigma : (\mathcal{B}_\sigma/=, \leq_\sigma) \to (\mathbf{KSN}(\mathcal{D}(\sigma)), \subseteq)$$

is an isomorphism, where $\mathbf{KSN}(D)$ *is the set of compact stable neighborhoods of D.*

Proof Any compact stable neighborhood of a coherent space D is a finite, pairwise disjoint, union of *prime open* sets of the form

$$\{ y \sqsupseteq x \mid y \in D \},$$

with $x \in D$ a finite element. By the completeness theorem, if

$$[\![\varphi]\!]_\sigma \cap [\![\varphi']\!]_\sigma = \emptyset$$

then $\vdash \varphi \wedge \varphi' = \mathbf{f}$, and $\varphi \vee \varphi' \in \mathcal{B}_\sigma$. Therefore, it is enough to show that for any prime open set $\{ y \sqsupseteq x \mid y \in D \}$, there is some $\varphi \in \mathcal{B}_\sigma$ such that

$$[\![\varphi]\!]_\sigma = \{ y \sqsupseteq x \mid y \in D \}.$$

However, finite elements of D are built up from the tokens. Hence it remains to show that for each token a of D there is an atomic assertion ψ, such that

$$[\![\psi]\!]_\sigma = \{ y \sqsupseteq \{ a \} \mid y \in D \}.$$

But this is just the conclusion of Proposition 9.1.

□

9.3 The Logic of DI

In this section we introduce a logic of **DI**. This framework is a generalization of the logic of $\mathbf{COH}_s$, derivable from the logic for $\mathbf{COH}_l$. For the same reason as mentioned at the beginning of the previous section, we use a disjunctive language which is formulated by using a mutual recursion of proof rules and syntactic rules. It appears that the logic of **DI** is more complicated due to the lifting type and the need to capture the prime assertions (Theorem 6.3, Definition 7.3).

As usual a meta-language of type expressions is given as follows:

$$\sigma ::= \mathbf{1} \mid \sigma + \tau \mid \sigma \times \tau \mid \sigma \rightarrow_s \tau \mid \sigma_\perp \mid t \mid rec\, t.\sigma$$

where t is a type variable and σ, τ ranges over type expressions. Note that we could have used the linear function space $\multimap$ and the exponential ! instead of $\rightarrow_s$ via the adjunction

$$\frac{!\sigma \multimap \tau}{\sigma \rightarrow_s \tau}.$$

But we made a choice not to deal with $\rightarrow_s$ directly this time.

Every closed type expression is interpreted as a dI-domain, with **1** as the one-point domain, $\times$ as the cartesian product, $+$ as the coalesced sum, $(\ )_\perp$ as lifting, and $\rightarrow_s$ as the stable function space. $rec\ t.\sigma$ the solution of prime information systems given in Chapter7. Write again $\mathcal{D}(\sigma)$ for the domain corresponding to σ.

For each type σ we introduce an assertion language C_σ according to the following rules.

Assertions

$$\mathbf{t}, \mathbf{f} : \sigma$$

$$\frac{\varphi,\ \psi : \sigma}{\varphi \wedge \psi : \sigma} \qquad \frac{\varphi \wedge \psi \leq \mathbf{f} \qquad \varphi,\ \psi : \sigma}{\varphi \vee \psi : \sigma}$$

$$\frac{\varphi : \sigma \qquad \psi : \tau}{\varphi \times \psi : \sigma \times \tau} \qquad \frac{\varphi : \sigma}{(\varphi)_\perp : \sigma_\perp}$$

$$\frac{\varphi : \sigma}{inl\varphi : \sigma + \tau} \qquad \frac{\psi : \tau}{inr\psi : \sigma + \tau}$$

$$\frac{\varphi : \sigma \qquad \psi : \tau}{\varphi \rightarrow \psi : \sigma \rightarrow_s \tau}$$

$$\frac{\varphi : \sigma[rec\, t.\sigma/t]}{\varphi : rec\, t.\sigma}$$

When we have $\varphi : \sigma$, we say φ is *well-formed*. Note that one of the rules,

$$\frac{\varphi \wedge \psi \leq \mathbf{f} \qquad \varphi,\ \psi : \sigma}{\varphi \vee \psi : \sigma},$$

makes use of the proof system to be formulated later. Here the situation is similar to the logic of coherent spaces, where a mutual recursion between syntactic rules and proof rules is allowed. The logic starts with some basic assertions, which are immediately justifiable to be well-formed. From these, one can use the proof rules to derive facts about them. Some of the facts, like $\varphi \wedge \psi \leq \mathbf{f}$, will allow one to form more assertions. In this way one get a disjunctive language.

Atomic assertions are some of the well-formed ones.

Atomic Assertions

$$\mathbf{A}_{\mathbf{1}_\perp}(\mathbf{t}_\perp) \qquad \frac{\mathbf{A}_{\sigma[rec\ t.\sigma/t]}(\varphi)}{\mathbf{A}_{rec\ t.\sigma}(\varphi)}$$

$$\frac{\mathbf{A}_\sigma(\varphi)}{\mathbf{A}_{\sigma+\tau}(\mathit{inl}\,\varphi)} \qquad \frac{\mathbf{A}_\tau(\psi)}{\mathbf{A}_{\sigma+\tau}(\mathit{inr}\,\psi)}$$

$$\frac{\mathbf{A}_\sigma(\varphi)}{\mathbf{A}_{\sigma_\perp}((\varphi)_\perp)}$$

$$\frac{\mathbf{A}_\sigma(\varphi)}{\mathbf{A}_{\sigma\times\tau}(\varphi\times\mathbf{t}_\tau)} \qquad \frac{\mathbf{A}_\tau(\psi)}{\mathbf{A}_{\sigma\times\tau}(\mathbf{t}_\sigma\times\psi)}$$

Moreover,

$\mathbf{A}_{\sigma\to_s\tau}(\bigwedge_{i\in I}\varphi_i\to\psi_i)$ if

- $\forall i\in I.\, \mathbf{P}_\sigma(\varphi_i)\ \&\ \mathbf{A}_\tau(\psi_i)$,
- $\exists k\in I.(\{\,\varphi_i \mid i\in I\,\}/\sim)\subseteq\lceil\varphi_k\rceil\ \&\ (\{\,\psi_i \mid i\in I\,\}/\sim)\subseteq[\![\psi_k]\!]$,
- $\forall i,\, j\in I.\, i\neq j\Rightarrow\neg(\varphi_i\to\psi_i\ \#\ \varphi_j\to\psi_j)$,
- $\forall j\in I\forall\chi\in[\![\psi_j]\!]\exists i\in I.\ \varphi_i\in\lceil\varphi_j\rceil\ \&\ \chi\sim\psi_i$.

Call φ *atomic* if $\mathbf{A}_\sigma(\varphi)$. It is intended that atomic assertions capture complete primes in the corresponding dI-domains. For this reason the specification for the atomic assertion of function space is a bit complicated; however this complication does not seem to be caused by the way the rules are given, but rather by the inherent complexity of the complete primes (see Theorem 6.3 and the discussion at the end of Section 7.4).

In the rules for atomic assertions several notations are used: they are $\mathbf{P}_\sigma$, $\#$, and $[\![\ \ ]\!]$. Some additional notations like $\lceil\ \ \rceil$ and $(\varphi)\downarrow$, although not needed right now, will be needed later in some other rules. So this is a good place to explain them all. The predicate $\mathbf{P}$ captures some assertions which correspond to prime open sets. They are called *prime assertions.*

The relation $\#$ captures *inconsistency* among the atomic assertions, so that from $\varphi \# \psi$ we can derive $\varphi \wedge \psi \leq \mathbf{f}$ by the propositional rule $(\#)$. The *atomizer*, $\llbracket\ \rrbracket$, gives a set $\llbracket\varphi\rrbracket$ of atomic assertions determined by a prime assertion φ. On the other hand, the *primer*, $\lceil\ \rceil$, gives a set $\lceil\varphi\rceil$ of derived prime assertions determined by a prime assertion φ.

Inconsistency Relation

$$\mathit{inl}\,\varphi \# \mathit{inr}\,\psi$$

$$\frac{\varphi \# \varphi'}{\mathit{inl}\,\varphi \# \mathit{inl}\,\varphi'} \qquad \frac{\psi \# \psi'}{\mathit{inr}\,\psi \# \mathit{inr}\,\psi'}$$

$$\frac{\varphi \# \varphi'}{(\varphi)_\perp \# (\varphi')_\perp}$$

$$\frac{\varphi \# \varphi'}{\varphi \times \mathbf{t} \# \varphi' \times \mathbf{t}} \qquad \frac{\psi \# \psi'}{\mathbf{t} \times \psi \# \mathbf{t} \times \psi'}$$

Note in the rules above all the assertions are assumed to be atomic.

The following are some additional rules for the inconsistency relation.

$$\frac{\mathbf{P}(\varphi \wedge \varphi') \quad \mathbf{A}(\psi) \quad \mathbf{A}(\psi') \quad \psi \# \psi'}{\varphi \to \psi \# \varphi' \to \psi'}$$

$$\frac{\mathbf{P}(\varphi \wedge \varphi') \quad \varphi \not\sim \varphi' \quad \mathbf{A}(\psi) \quad \mathbf{A}(\psi') \quad \psi \sim \psi'}{\varphi \to \psi \# \varphi' \to \psi'}$$

$$\frac{\mathbf{A}(\bigwedge_{i\in I} \varphi_i \to \psi_i) \quad \mathbf{A}(\bigwedge_{j\in J} \varphi'_j \to \psi'_j) \quad \exists i \in I \exists j \in J.\, \varphi_i \to \psi_i \# \varphi'_j \to \psi'_j}{\bigwedge_{i\in I} \varphi_i \to \psi_i \# \bigwedge_{j\in J} \varphi'_j \to \psi'_j}$$

Here the *similarity relation* $\sim$ is a relation on atomic assertions (they can be easily extended to prime assertions) which is very close to the *syntactic equality*. The only difference between $\sim$ and syntactic equality arises from the fact that for atomic assertions of function space, we ignore the order of the conjuncts. More precisely, let

$$\begin{aligned}
&(\varphi_\perp)^* = (\varphi^*)_\perp,\\
&(\varphi \times \psi)^* = (\varphi^*) \times (\psi^*),\\
&(\mathit{inl}\,\varphi)^* = \mathit{inl}(\varphi^*),\\
&(\mathit{inr}\,\psi)^* = \mathit{inr}(\psi^*),\\
&(\varphi \to \psi)^* = (\varphi)^* \to (\psi)^*,\\
&(\textstyle\bigwedge_{i\in I} \varphi_i)^* = \bigcup_{i\in I}\{(\varphi_i)^*\}.
\end{aligned}$$

Here assertions are assumed atomic. Then $\sim$ can be formally defined by the following rules. All the assertions in the following table are assumed to be atomic.

Similarity Relation

$$\mathbf{t}_\perp \sim \mathbf{t}_\perp$$

$$\frac{\varphi \sim \varphi'}{\mathit{inl}\,\varphi \sim \mathit{inl}\,\varphi'}$$

$$\frac{\psi \sim \psi'}{\mathit{inr}\,\psi \sim \mathit{inr}\,\psi'}$$

$$\frac{\varphi \sim \varphi'}{(\varphi)_\perp \sim (\varphi')_\perp}$$

$$\frac{\varphi \sim \varphi'}{\varphi \times \mathbf{t} \sim \varphi' \times \mathbf{t}}$$

$$\frac{\psi \sim \psi'}{\mathbf{t} \times \psi \sim \mathbf{t} \times \psi'}$$

$$\frac{\{\, (\varphi_i \to \psi_i)^* \mid i \in I \,\} = \{\, (\varphi'_j \to \psi'_j)^* \mid j \in J \,\}}{\bigwedge_{i\in I} \varphi_i \to \psi_i \;\sim\; \bigwedge_{j\in J} \varphi'_j \to \psi'_j}$$

The similarity relation $\sim$ can then be easily extended to prime assertions by defining

$$\bigwedge_{i\in I} \varphi_i \sim \bigwedge_{j\in J} \psi_j \iff (\bigwedge_{i\in I} \varphi_i)^* = (\bigwedge_{j\in J} \psi_j)^*.$$

The similarity relation is introduced as a simple way to decide whether two atomic assertions have the same interpretation as stable neighborhoods. It will be shown later (Proposition 9.7) that two prime (specially atomic) assertions are similar if and only if they have the same interpretation as stable neighborhoods. The following are the definition for the other notations.

$$\mathbf{P}_\sigma(\varphi) \iff \varphi \equiv \bigwedge_{i\in I} \varphi_i \;\&\; \forall i \in I.\, \mathbf{A}_\sigma(\varphi_i) \;\&\; \forall i \neq j.\, \neg(\varphi_i \# \varphi_j),$$
$$(\varphi)\!\downarrow \iff \mathbf{P}(\varphi) \;\&\; \varphi \equiv \bigwedge_{i\in I} \varphi_i \;\&\; I \neq \emptyset$$

Let φ be a prime assertion, i.e. $\mathbf{P}(\varphi)$. Then there are assertions φ_i, such that $\varphi \equiv \bigwedge_{i\in I} \varphi_i$ and $\forall i \in I.\, \mathbf{A}(\varphi_i)$. We define

$$\lceil \varphi \rceil = \{\, \bigwedge_{j\in J} \varphi_j \mid \mathbf{P}(\bigwedge_{j\in J} \varphi_j) \;\&\; J \subseteq I \,\}/\!\sim$$

and

$$\llbracket \varphi \rrbracket = \{\, \bigwedge_{j\in J} \varphi_j \mid \mathbf{A}(\bigwedge_{j\in J} \varphi_j) \;\&\; J \subseteq I \,\}/\!\sim.$$

Note that when forming the sets $\lceil \varphi \rceil$ and $\llbracket \varphi \rrbracket$ we took the quotient over $\sim$. Thus it is reminded that equality of sets $\llbracket \varphi \rrbracket = \llbracket \varphi' \rrbracket$ is the equality over the quotient sets. Note also that if we use ! and $\multimap$ instead of $\to_s$ for the type expressions, the simple relation

$$\llbracket !\varphi \rrbracket_{!\sigma} = \{\, !\psi \mid \psi \in \lceil \varphi \rceil_\sigma \,\}$$

will hold.

It follows from the definition of **P** that

$\mathbf{P}_{\sigma\to_s\tau}(\bigwedge_{i\in I} \varphi_i \to \psi_i)$ if and only if
- $\forall i \in I.\, \mathbf{P}_\sigma(\varphi_i)\ \&\ \mathbf{A}_\tau(\psi_i)$,
- $\forall i, j \in I.\, i \neq j \Rightarrow \neg(\varphi_i \to \psi_i \# \varphi_j \to \psi_j)$, and
- $\forall j \in I \forall \chi \in \llbracket \psi_j \rrbracket \exists i \in I.\, \varphi_i \in \lceil \varphi_j \rceil\ \&\ \chi \sim \psi_i$.

Comparing this with the definition for **A**, here we do not require that

$$\exists k \in I.\, (\{\, \varphi_i \mid i \in I \,\}/{\sim}) \subseteq \lceil \varphi_k \rceil\ \&\ (\{\, \psi_i \mid i \in I \,\}/{\sim}) \subseteq \llbracket \psi_k \rrbracket.$$

The propositional rules remain unchanged.

Propositional Rules

(t)	$\varphi \leq \mathbf{t}$	(f)	$\mathbf{f} \leq \varphi$
(Ref)	$\varphi \leq \varphi$	(#)	$\dfrac{\varphi \# \psi}{\varphi \wedge \psi \leq \mathbf{f}}$
(Trans)	$\dfrac{\varphi \leq \varphi' \quad \varphi' \leq \varphi''}{\varphi \leq \varphi''}$		
$(\leq - =)$	$\dfrac{\varphi \leq \psi \quad \psi \leq \varphi}{\varphi = \psi}$		
$(= - \leq)$	$\dfrac{\varphi = \varphi'}{\varphi \leq \varphi'} \qquad \dfrac{\varphi = \varphi'}{\varphi' \leq \varphi}$		
$(\wedge - \leq)$	$\varphi \wedge \varphi' \leq \varphi \qquad \varphi \wedge \varphi' \leq \varphi'$		
$(\leq - \wedge)$	$\dfrac{\varphi \leq \varphi' \quad \varphi \leq \varphi''}{\varphi \leq \varphi' \wedge \varphi''}$		
$(\vee - \leq)$	$\dfrac{\varphi \leq \varphi' \quad \psi \leq \varphi'}{\varphi \vee \psi \leq \varphi'}$		
$(\leq - \vee)$	$\varphi \leq \varphi' \vee \varphi \qquad \varphi' \leq \varphi' \vee \varphi$		
$(\wedge - \vee)$	$\varphi \wedge (\varphi_1 \vee \varphi_2) \leq (\varphi \wedge \varphi_1) \vee (\varphi \wedge \varphi_2)$		

Assertions in the above table are assumed to be all well-formed. The proof system consists of several groups of axioms and rules given below. There are type-specific rules which provide relationships between axioms of different types. There are also axioms that tell us how logical constructions interact with type constructions. We have encountered similar sets of rules many times so this time no explanation is needed.

Sum

$$(\mathit{inl}-\leq)\quad \frac{\varphi \leq \psi}{\mathit{inl}\,\varphi \leq \mathit{inl}\,\psi}$$

$$(\mathit{inr}-\leq)\quad \frac{\varphi \leq \psi}{\mathit{inr}\,\varphi \leq \mathit{inr}\,\psi}$$

$$(\mathit{inl}-\wedge)\quad \mathit{inl}\,(\bigwedge_{i\in I}\varphi_i) = \bigwedge_{i\in I}\mathit{inl}\,\varphi_i$$

$$(\mathit{inr}-\wedge)\quad \mathit{inr}\,(\bigwedge_{i\in I}\psi_i) = \bigwedge_{i\in I}\mathit{inr}\,\psi_i$$

$$(\mathit{inl}-\vee)\quad \mathit{inl}\,(\bigvee_{i\in I}\varphi_i) = \bigvee_{i\in I}\mathit{inl}\,\varphi_i$$

$$(\mathit{inr}-\vee)\quad \mathit{inr}\,(\bigvee_{i\in I}\psi_i) = \bigvee_{i\in I}\mathit{inr}\,\psi_i$$

Product

$$(\times-\leq)\quad \frac{\psi \leq \psi' \quad \varphi \leq \varphi'}{\psi \times \varphi \leq \psi' \times \varphi'}$$

$$(\times-\vee)\quad \varphi \times (\bigvee_{i\in I}\psi_i) = \bigvee_{i\in I}(\varphi \times \psi_i)$$

$$(\bigvee_{i\in I}\varphi_i) \times \psi = \bigvee_{i\in I}(\varphi_i \times \psi)$$

$$(\times-\wedge)\quad \bigwedge_{i\in I}(\varphi_i \times \psi_i) = (\bigwedge_{i\in I}\varphi_i) \times (\bigwedge_{i\in I}\psi_i)$$

Function Space

$$(\rightarrow-\mathbf{f})\quad \frac{\varphi\downarrow}{(\varphi \rightarrow \mathbf{t}) \leq \mathbf{f}}$$

$$(\rightarrow)\quad \frac{\lceil\varphi\rceil \neq \lceil\varphi'\rceil \qquad \mathbf{P}(\varphi \wedge \varphi') \qquad \mathbf{P}(\psi)}{(\varphi \rightarrow \psi) \wedge (\varphi' \rightarrow \psi) \leq \mathbf{f}}$$

$$(\rightarrow-\leq)\quad \frac{\chi \in \lceil\psi\rceil \qquad \mathbf{P}(\varphi)}{\varphi \rightarrow \psi \leq \bigvee_{\xi\in\lceil\varphi\rceil}(\xi \rightarrow \chi)}$$

$$(\rightarrow-\wedge)\quad \bigwedge_{i\in I}(\varphi_i \rightarrow \psi_i) \leq (\bigwedge_{i\in I}\varphi_i) \rightarrow (\bigwedge_{i\in I}\psi_i)$$

$$(\rightarrow-\vee)\quad \frac{\mathbf{P}(\varphi)}{\varphi \rightarrow (\bigvee_{i\in I}\psi_i) = \bigvee_{i\in I}(\varphi \rightarrow \psi_i)}$$

$$(\bigvee_{i\in I}\varphi_i) \rightarrow \psi = \bigwedge_{i\in I}(\varphi_i \rightarrow \psi)$$

We have the rule $(\rightarrow -\mathbf{f})$ because for any stable function $f : D \rightarrow E$, $\mu\, f^{-1}(E) = \bot_D$ (Definition 8.4). Therefore if φ converges it is not possible for $\mu\,[\![\varphi]\!] \subseteq \mu\, f^{-1}([\![\mathbf{t}]\!])$ to hold. $[\![\varphi \rightarrow \mathbf{t}]\!]$ must be empty by Definition 8.5. This is very different form what we used to have with Scott continuous functions. The second rule, $(\rightarrow)$, reflects the second condition in Theorem 6.3. If that condition is not met, it defines no stable function. The third rule is a close translation of Proposition 8.7.

There is nothing unusual about the rules for lifting.

Lifting

$$(\bot - \leq) \quad \frac{\varphi \leq \psi}{\varphi_\bot \leq \psi_\bot}$$

$$(\bot - \wedge) \quad (\varphi_1 \wedge \varphi_2)_\bot = (\varphi_1)_\bot \wedge (\varphi_2)_\bot$$

$$(\bot - \vee) \quad (\bigvee_{i \in I} \varphi_i)_\bot = \bigvee_{i \in I} (\varphi_i)_\bot$$

Now we give an interpretation (semantics) for assertions. For each closed type expression σ we define an *interpretation function*

$$[\![\ ]\!]_\sigma : C_\sigma \rightarrow \mathbf{KSN}(\mathcal{D}(\sigma))$$

with $\mathbf{KSN}(D)$ the collection of compact stable neighborhoods of D. $[\![\ ]\!]_\sigma$ is defined in the following structural way.

For each closed type expression σ, we define

$$\begin{aligned} &[\![\mathbf{t}]\!]_\sigma = \mathcal{D}(\sigma), \\ &[\![\mathbf{f}]\!]_\sigma = \emptyset, \\ &[\![\varphi \vee \psi]\!]_\sigma = [\![\varphi]\!]_\sigma \cup [\![\psi]\!]_\sigma, \\ &[\![\varphi \wedge \psi]\!]_\sigma = [\![\varphi]\!]_\sigma \cap [\![\psi]\!]_\sigma. \end{aligned}$$

With respect to type constructions we define

$$\begin{aligned} &[\![\varphi \times \psi]\!]_{\sigma \times \tau} = \{ (u, v) \mid u \in [\![\varphi]\!]_\sigma \ \&\ v \in [\![\psi]\!]_\tau \}, \\ &[\![\mathit{inl}\, \varphi]\!]_{\sigma + \tau} = \{ (0, u) \mid u \in [\![\varphi]\!]_\sigma \setminus \{ \bot_{\mathcal{D}(\sigma)} \} \} \\ &\qquad \cup \{ x \in \mathcal{D}(\sigma + \tau) \mid \bot_{\mathcal{D}(\sigma)} \in [\![\varphi]\!]_\sigma \}, \\ &[\![\mathit{inr}\, \varphi]\!]_{\sigma + \tau} = \{ (1, u) \mid u \in [\![\varphi]\!]_\tau \setminus \{ \bot_{\mathcal{D}(\tau)} \} \} \\ &\qquad \cup \{ x \in \mathcal{D}(\sigma + \tau) \mid \bot_{\mathcal{D}(\tau)} \in [\![\varphi]\!]_\tau \}, \\ &[\![\varphi \rightarrow \psi]\!]_{\sigma \rightarrow_s \tau} = \{ f \in \mathcal{D}(\sigma) \rightarrow_s \mathcal{D}(\tau) \mid [\![\varphi]\!]_\sigma \sqsubseteq_\mu f^{-1}([\![\psi]\!]_\tau) \}, \\ &[\![(\varphi)_\bot]\!]_{(\sigma)_\bot} = \{ (0, u) \mid u \in [\![\varphi]\!]_\sigma \}, \\ &[\![\varphi]\!]_{rect.\sigma} = \{ \epsilon_\sigma(u) \mid u \in [\![\varphi]\!]_{\sigma[(rect.\sigma)\backslash t]} \}. \end{aligned}$$

where $\epsilon_\sigma : \mathcal{D}(\sigma[(rect.\sigma)\backslash t]) \rightarrow \mathcal{D}(rect.\sigma)$ is the equality arising form the recursively defined stable domains.

9.4 Completeness

In this section we show that the proof system for **DI** is sound and complete. The definitions, propositions, and proofs have a similar style to those in Section 9.4.

Definition 9.3 *For $\varphi, \psi \in C_\sigma$, write $\models_\sigma \varphi \leq_\sigma \psi$ if $[\![\varphi]\!]_\sigma \subseteq [\![\psi]\!]_\sigma$. Write $\vdash_\sigma \varphi \leq_\sigma \psi$ if $\varphi \leq_\sigma \psi$ can be derived from the proof system given in the previous section. The proof system is called sound if $\vdash \varphi \leq_\sigma \psi$ implies $\models \varphi \leq_\sigma \psi$. It is complete if $\models \varphi \leq_\sigma \psi$ implies $\vdash \varphi \leq_\sigma \psi$. An axiom is valid if it is a valid formula. A rule is sound if it produces valid formulae from valid formulae.*

Proposition 9.6 *There is an isomorphism between atomic assertions of σ and complete primes of $\mathcal{D}(\sigma)$ such that for any φ, ψ, atomic assertions of type σ, $\varphi \# \psi$ if and only if*

$$[\![\varphi]\!]_\sigma \cap [\![\psi]\!]_\sigma \subseteq \emptyset.$$

Proposition 9.7 *Let φ, ψ be prime assertions of type σ. Then $\varphi \sim \psi$ if and only if $[\![\varphi]\!] = [\![\psi]\!]$.*

Proof Only if: By inspecting the definitions.

If: Note that for prime assertions φ and ψ, $(\varphi)^* \neq (\psi)^*$ implies $[\![\varphi]\!] \neq [\![\psi]\!]$ provided that we have $\varphi \not\sim \psi$ implies $[\![\varphi]\!] \neq [\![\psi]\!]$ for atomic assertions φ and ψ. Therefore it is enough to check the conclusion for atomic assertions. We show that $\varphi \not\sim \psi$ implies $[\![\varphi]\!] \neq [\![\psi]\!]$. This is done by structural induction on types. For the base type **1**, $\varphi : \mathbf{1}$ is prime if and only if $\varphi \equiv \bigwedge \emptyset$ by definition, since type **1** has no atomic assertion. Therefore the statement 'for prime assertions φ, ψ, $\varphi \not\sim \psi$ implies $[\![\varphi]\!] \neq [\![\psi]\!]$' is vacuously true. Next we show that the rules for the similarity relation preserves the property $\varphi \not\sim \psi$ implies $[\![\varphi]\!] \neq [\![\psi]\!]$, but that can be done by inspecting each rule.

□

Proposition 9.8 *1. Suppose φ is an atomic assertion of type σ. Then there exists $p \in \mathcal{D}(\sigma)$, a complete prime, such that $[\![\varphi]\!]_\sigma = \uparrow x$. 2. Suppose $\varphi \in C_\sigma$ and $\mathbf{P}(\varphi)$. Then there exists a finite element $x \in \mathcal{D}(\sigma)$ such that $[\![\varphi]\!]_\sigma = \uparrow x$. 3. Let φ be a prime assertion of σ and write $\check{\varphi}$ for the finite* *element of $\mathcal{D}(\sigma)$ such that $[\![\varphi]\!]_\sigma = \uparrow \breve{\varphi}$. If $\lceil \varphi \rceil = \{ \varphi_i \mid i \in I \}$, then* $\{ \breve{\varphi}_i \mid i \in I \} = \{ y \in (\mathcal{D}(\sigma))^0 \mid y \sqsubseteq \breve{\varphi} \}$; *If $\lceil \varphi \rceil = \{ \varphi_i \mid i \in I \}$, then* $\{ \breve{\varphi}_i \mid i \in I \} = \{ y \in (\mathcal{D}(\sigma))^p \mid y \sqsubseteq \breve{\varphi} \}$. *4. Suppose $\varphi \in \mathcal{C}_\sigma$ and $\varphi \! \downarrow$. Then* $\mathcal{D}_\sigma \nsubseteq [\![\varphi]\!]_\sigma$.

Recall that D^0 stands for the set of finite elements of D and D^p stands for the set of complete primes of D.

Proof The proofs for 3 and 4 are routine. We use an induction on types to show 1 and 2. The nontrivial case is function space. However, that follows from the properties of stable joinable sets given in Theorem 6.3.

$\square$

Proposition 9.9 *If $\lceil \varphi \rceil \neq \lceil \psi \rceil$ for prime assertions φ, ψ of σ, then* $[\![\varphi]\!]_\sigma \neq [\![\psi]\!]_\sigma$.

Proof It is easy to show that $\lceil \varphi \rceil \neq \lceil \psi \rceil$ implies $\varphi \not\sim \psi$. The conclusion then follows from Proposition 9.7.

$\square$

Theorem 9.4 establishes the soundness of the proof system.

Theorem 9.4

- *The logical axioms are valid and logical rules sound.*
- *The axioms for sum, product, lifting, and function space are valid.*
- *The rules for sum, product, lifting, and function space are sound.*

Proof We check this fact for the function space construction. Other cases are much easier.

The soundness of $(\rightarrow - \vee)$, $(\rightarrow - \wedge)$ follows from Proposition 8.6 and Proposition 9.8. The soundness of $(\rightarrow - \leq)$ follows from Proposition 8.7, Proposition 9.7, and Proposition 9.8, while the soundness of $(\rightarrow - \mathbf{f})$ follows from Definition 8.5 and the fourth conclusion of Proposition 9.8. The rule $(\rightarrow)$ is sound since if f is a stable function and $x, y \in \mu f^{-1}(A)$ with $x \uparrow y$, then we must have $x = y$.

$\square$

Definition 9.4 *Write $\mathcal{Q}_\sigma$ for the proof system associated with type σ. $\mathcal{Q}_\sigma$ is called prime complete if it has property p_0, prime normal if it has property*

p_1, and complete if it has property p_2, where

$$(p_0)\qquad \forall\varphi,\psi:\sigma.\,(\mathbf{P}(\varphi)\ \&\ \mathbf{P}(\psi)\ \&\ [\![\varphi]\!]_\sigma\subseteq[\![\psi]\!]_\sigma\Longrightarrow\vdash\varphi\le\psi),$$

$$(p_1)\qquad \varphi:\sigma\Longrightarrow\exists\{\varphi_i\mid i\in I\}.\,\forall i\in I.\,\mathbf{P}(\varphi_i)\ \&\ \vdash\varphi=\bigvee_{i\in I}\varphi_i,$$

$$(p_2)\qquad \forall\varphi,\psi:\sigma.\,([\![\varphi]\!]\subseteq[\![\psi]\!]\Longrightarrow\vdash\varphi\le\psi).$$

In (p_1), $\bigvee_{i\in I}\varphi_i$ is called a prime normal form of φ.

Clearly $\mathcal{Q}_1$ has property p_0, p_1, and p_2. The proof for the completeness of the system is achieved by showing that each type construction preserves property (p_0), (p_1), and (p_2). We omit the cases for lifting, sum, and product, since they are straightforward. Note that for each case the proof of (p_2) is routine. It follows from (p_0) and (p_1) directly.

Proposition 9.10 *Function space preserves p_0, p_1, and p_2.*

Proof The proof of property p_0 is easy. We prove the interesting property p_1 for function space. Similarly to the proof of Proposition 9.5 we have

$$\vdash\varphi\to\psi=\bigvee_{\kappa:\lceil\psi\rceil\to\lceil\varphi\rceil\text{ onto}}\ \bigwedge_{a\in\lceil\psi\rceil}(\kappa(a)\to a).$$

Note

$$\bigwedge_{a\in\lceil\psi\rceil}(\kappa(a)\to a)$$

need not be a prime assertion, since the property

$$\forall j\in I\,\forall\xi\in\lceil\psi_j\rceil\,\exists i\in I.\ \varphi_i\in\lceil\varphi_j\rceil\ \&\ \xi\sim\psi_i$$

required for a prime assertion $\bigwedge_{i\in I}\varphi_i\to\psi_i$ may not hold. To get a prime normal form we have to apply the same procedure again to $\kappa(a)\to a$ for each conjuncts of

$$\bigwedge_{a\in\lceil\psi\rceil}(\kappa(a)\to a).$$

Repeat this procedure for a finite number of times we get a prime normal form for $\varphi\to\psi$.

□

From the above we get

Theorem 9.5 *The proof system for the logic of coherent spaces is complete.*

The expressive result is what one can expect, and the proof is similar to that of Theorem 9.3.

Theorem 9.6 *Let σ be a closed type expression. Then*

$$[\![\ ]\!]_\sigma : (\mathcal{C}_\sigma/=,\ \leq_\sigma) \rightarrow (\mathbf{KSN}(\mathcal{D}(\sigma)),\ \subseteq)$$

is an isomorphism, where $\mathbf{KSN}(D)$ *is the set of compact stable neighborhoods of D.*

9.5 An Alternative Approach

There is no absolute necessity to have disjunctive assertions. A trade-off is involved here: if the assertions are disjunctive, then the proof rules are simpler. However, if the assertions are chosen to be conventional (like those for the logic of SFP domains), some of the proof rules may become clumsy for the stable case. Nevertheless, in order not to commit ourselves too early to the disjunctive treatment of assertions, this section discusses an alternative, using a conventional assertion language for the proof system.

The entailment should be interpreted as set-inclusion, not the minimal relation $\sqsubseteq_\mu$, on stable neighborhoods. This is because, as stated in Section 1.4, we want the denotational semantics to agree with the axiomatic semantics. In other words, the partial order of information on the computations should be determined by the containment of properties:

$$a \sqsubseteq b \Longleftrightarrow \{A \mid a \models A\} \subseteq \{B \mid b \models B\}.$$

The elements of the domain should be recovered as some complete theories x which are consistent and closed under the entailment. These make the minimal relation $\sqsubseteq_\mu$ inappropriate for the interpretation of the entailment. One of the reasons is that we should always have $\varphi \leq \mathbf{t}$ for any assertion φ.

It all boils down to the question of how to interpret the logical connectives $\wedge$ and $\vee$ in the stable neighborhoods $(\mathbf{SN}(D),\ \subseteq)$. There are two possiblities. The first, interpreting them as set union and intersection, leads to the disjunctive logic we have dealt with already. The other possibility is to interpret them as $\sqcap$ and $\sqcup$ in the lattice $(\mathbf{SN}(D),\ \subseteq)$, which we deal with now.

Let's consider the same language of type expressions for coherent spaces:

$$\sigma ::= O \mid \sigma \times \tau \mid \sigma \otimes \tau \mid \sigma + \tau \mid \sigma \multimap \tau \mid !\sigma \mid t \mid rec\, t.\sigma.$$

The specifications of atomic assertions and atomic inconsistency remain the same. However, assertions are now generated freely, without the disjunctive restriction:

Assertions

$$\mathbf{t}, \mathbf{f} : \sigma \qquad \frac{\mathbf{A}_\sigma(\varphi)}{\varphi : \sigma}$$

$$\frac{\varphi,\ \psi : \sigma}{\varphi \wedge \psi : \sigma} \qquad \frac{\varphi,\ \psi : \sigma}{\varphi \vee \psi : \sigma}$$

$$\frac{\varphi : \sigma \qquad \psi : \tau}{\varphi \times \psi : \sigma \times \tau} \qquad \frac{\varphi : \sigma \qquad \psi : \tau}{\varphi \otimes \psi : \sigma \otimes \tau}$$

$$\frac{\varphi : \sigma}{inl\,\varphi : \sigma + \tau} \qquad \frac{\psi : \tau}{inr\,\psi : \sigma + \tau}$$

$$\frac{\varphi : \sigma}{!\varphi : !\sigma} \qquad \frac{\varphi : \sigma \qquad \psi : \tau}{\varphi \multimap \psi : \sigma \multimap \tau}$$

$$\frac{\varphi : \sigma[rec\, t.\sigma/t]}{\varphi : rec\, t.\sigma}$$

The interpretation of assertions is given differently than before. For each closed type expression σ, let

$$\begin{aligned}
&[\![\mathbf{t}]\!]_\sigma = \mathcal{D}(\sigma),\\
&[\![\mathbf{f}]\!]_\sigma = \emptyset,\\
&[\![\top]\!]_O = \top,\\
&[\![\varphi \vee \psi]\!]_\sigma = [\![\varphi]\!]_\sigma \sqcup [\![\psi]\!]_\sigma,\\
&[\![\varphi \wedge \psi]\!]_\sigma = [\![\varphi]\!]_\sigma \sqcap [\![\psi]\!]_\sigma.
\end{aligned}$$

Note the change occurred in

$$[\![\varphi \vee \psi]\!]_\sigma = [\![\varphi]\!]_\sigma \sqcup [\![\psi]\!]_\sigma,$$

where the logical disjunction is interpreted as the least upper bound in the lattice of stable neighborhoods, instead of set union. Since finite intersection preserves stable neighborhoods, $\sqcap$ is the same as $\cap$. The interpretation for type constructions, however, remain the same.

With respect to proof rules, one should be careful that not all propositional rules are valid any more. With the logical disjunction interpreted as the least upper bound, the distributive law does not hold. We do not have

$$\varphi \wedge (\psi_1 \vee \psi_2) \leq (\varphi \wedge \psi_1) \vee (\varphi \wedge \psi_2)$$

because of Example 8.2, showing that $(\mathbf{SN}(D), \subseteq)$ need not be distributive.

To get around this rather unpleasant situation, a slightly modified rule like

$$\frac{\psi_1 \wedge \psi_2 \leq \mathbf{f}}{\varphi \wedge (\psi_1 \vee \psi_2) \leq (\varphi \wedge \psi_1) \vee (\varphi \wedge \psi_2)}$$

is essential. However, one can feel that this is suggesting the use of disjunctive assertions!

Indeed, it seems that although the assertions are not required to be disjunctive, the disjunctive property still has much to do with the new proof system. When it comes to the proof of completeness, a disjunctive prime normal form for assertions has to be used (by a disjunctive prime normal form we mean a disjunction $\bigvee_{i \in I} \varphi_i$ such that every φ_i is prime, and

$$\varphi_i \wedge \varphi_j \leq \mathbf{f}$$

for $i \neq j$). To illustrate this, let's consider the proof system for product. Note that all the existing rules for product remain sound and hence are adopted as they stand. However they are not complete. Consider the type expression $O \times O$ and an assertion

$$\mathbf{t} \times \top \vee \top \times \mathbf{t}.$$

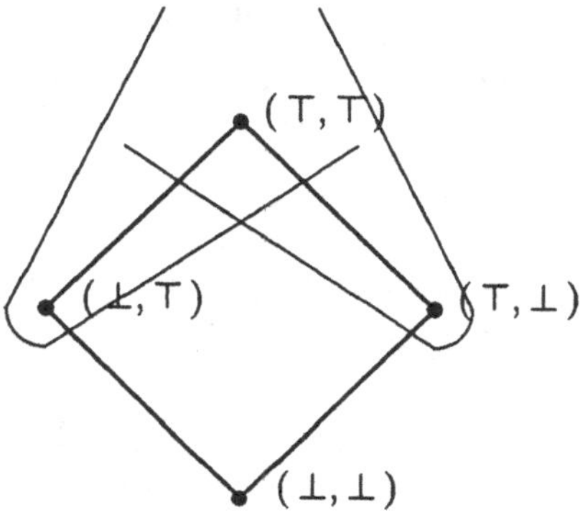

From the picture it is clear that we should have

$$\mathbf{t} \times \top \vee \top \times \mathbf{t} = \mathbf{t} \times \mathbf{t}.$$

But this cannot be derived without introducing a new rule. Introducing

$$\bigvee_{i\in I} \varphi_i \times \psi_i = (\bigvee_{i\in I} \varphi_i) \times (\bigvee_{i\in I} \psi_i)$$

would have solved the problem for the example, only that it is unsound.

After some deliberation we can see that the following new rule can be introduced:

$$(\times - \mathbf{P}) \qquad \frac{\mathbf{P}(\bigwedge_{i\in I} \varphi_i) \qquad \mathbf{P}(\bigwedge_{i\in I} \psi_i)}{\bigvee_{i\in I} \varphi_i \times \psi_i = (\bigvee_{i\in I} \varphi_i) \times (\bigvee_{i\in I} \psi_i)}.$$

With this rule it is possible to show that the proof system for product is sound and complete, or, more precisely, it preserves sound and complete proof systems, via a disjunctive prime normal form for assertions.

It is useful to note that if $\bigvee_{i\in I} \varphi_i$ is a disjunction with $\bigwedge_{i\in I} \varphi_i$ prime, then there is a single prime assertion ψ such that

$$\bigvee_{i\in I} \varphi_i = \psi$$

is derivable.

Theorem 9.7 *The proof system for product, consisting of* $(\times - \leq)$, $(\times - \vee)$, $(\times - \wedge)$, *and* $(\times - \mathbf{P})$ *preserves sound and complete proof systems.*

Proof For soundness it is enough to show that $(\times - \mathbf{P})$ is sound. Note $\mathbf{P}(\bigwedge_{i\in I} \varphi_i)$ and $\mathbf{P}(\bigwedge_{i\in I} \psi_i)$ implies that all φ_i's and ψ_i's are prime. Prime assertions capture prime open sets. Therefore there are a_i's and b_i's such that

$$[\![\varphi_i]\!] = \uparrow a_i, \text{and } [\![\psi_i]\!] = \uparrow b_i$$

for $i \in I$. The fact that $\bigwedge_{i\in I} \varphi_i$ and $\bigwedge_{i\in I} \psi_i$ are themselves prime implies that $\{a_i \mid i \in I\}$ and $\{b_i \mid i \in I\}$ are compatible sets. To show

$$[\![\bigvee_{i\in I} \varphi_i \times \psi_i]\!] = [\![(\bigvee_{i\in I} \varphi_i) \times (\bigvee_{i\in I} \psi_i)]\!],$$

it is enough to show that (Proposition 8.1 is relevant here) for any $i, j \in I$,

$$(a_i, b_j) \in [\![\bigvee_{i\in I} \varphi_i \times \psi_i]\!].$$

This is done as follows. We have

$$(a_i, b_i) \in [\![\bigvee_{i \in I} \varphi_i \times \psi_i]\!]$$

and

$$(a_j, b_j) \in [\![\bigvee_{i \in I} \varphi_i \times \psi_i]\!].$$

But $a_i \uparrow a_j$, $b_i \uparrow b_j$. Therefore $(a_i, b_i) \uparrow (a_j, b_j)$. That means

$$(a_i, b_i) \sqcap (a_j, b_j) \in [\![\bigvee_{i \in I} \varphi_i \times \psi_i]\!]$$

since $[\![\bigvee_{i \in I} \varphi_i \times \psi_i]\!]$ is a stable neighborhood. Noticing that

$$(a_i, b_i) \sqcap (a_j, b_j) = (a_i \sqcap a_j,\ b_i \sqcap b_j),$$

we conclude that

$$(a_i, b_j) \in [\![\bigvee_{i \in I} \varphi_i \times \psi_i]\!].$$

The key to completeness is the existence of disjunctive prime normal forms. We show that product preserves the existence of disjunctive prime normal forms.

Let $\bigvee_{i \in I} \varphi_i \times \psi_i$ be an assertion of the product, where φ_i's and ψ_i's are prime. For some n and I_j's such that $I_j \subseteq I$ for $1 \leq k \leq n$, we have

$$\begin{aligned} &\bigvee_{i \in I} \varphi_i \times \psi_i \\ &= \bigvee_{1 \leq j \leq n} (\bigvee_{k \in I_j} \varphi_k \times \psi_k), \end{aligned}$$

where each I_j is a largest subset of I which makes $\bigvee_{k \in I_j} \varphi_k \times \psi_k$ a prime assertion (adding one more index to I_j would make the assertion inconsistent). Now applying $(\times - \mathbf{P})$ to each disjuncts $\bigvee_{k \in I_j} \varphi_k \times \psi_k$, we get

$$\begin{aligned} &\bigvee_{i \in I} \varphi_i \times \psi_i \\ &= \bigvee_{1 \leq j \leq n} (\bigvee_{k \in I_j} \varphi_k) \times (\bigvee_{k \in I_j} \psi_k). \end{aligned}$$

As pointed out earlier,

$$\begin{aligned} &(\bigvee_{k \in I_j} \varphi_k) \times (\bigvee_{k \in I_j} \psi_k) \\ &= \alpha_j \times \beta_j, \end{aligned}$$

with $\alpha_j \times \beta_j$ a prime assertion. Therefore

$$\begin{aligned} & \bigvee_{i \in I} \varphi_i \times \psi_i \\ & = \bigvee_{1 \leq j \leq n} \alpha_j \times \beta_j \end{aligned}$$

a disjunctive prime normal form.

□

A similar property holds for tensor product, linear function space, and exponential, with a few new rules introduced. However, even though the new rules are bearable, the proofs for soundness and completeness become much more involved, not to mention what has to be done for dI-domains!

For this reason, the details for these types are not included here.

Chapter 10

Research Topics

We have studied domain logic of two prominent categories for denotational semantics: the SFP domains and the stable domains. The promise of domain logic was first adequately demonstrated in Abramsky's work 'Domain Theory in Logical Form' ([Ab87]). Although this monograph definitely adds more to our appreciation of the importance, this should really be counted as a side effect. The purposes of the monograph is, as pointed out at the beginning, to convey to the reader the state of knowledge on logic of domains, and to point to areas where more research is to be done. We hope that we have achieved the first by presenting some of the recent results on a variety of topics about important aspects of logic of domains.

As for the second purpose of the monograph, there are many places where we have explicitly mentioned further work along the way when the materials were presented. However, a more systematic disclosure of the research topics directly follow after this monograph is certainly more helpful, and that is what we are going to do now. Note that the following list is not meant to be exhaustive. It is inevitable some of the important research topics may have not been mentioned here.

1. In Chapter 3 we presented a category of strongly finite sequent structures to represent SFP domains. Not much is done with the more basic sequent structures. Although there is some understanding about what kind of cpos the sequent structures represent ([DG90a]), there is still no categorical treatment of them. It appears non-trivial to find the right definition of morphisms on sequent structures. Once the morphisms are found,

one can further discuss constructions and recursively defined structures.

2. It is a pity the connection of Hoare logic for while-programs and the semantics generated logic has not been fully explored. It seems something interesting may be happening here, especially about the completeness issue, as pointed out in Section 4.5.

3. As to Brookes' proof system, although the assertion language is shown to be derivable from Plotkin's domain of resumptions, it would be nice to show that the proof rules of Brookes' proof system are also semantics derived.

4. Chapter 5 provided a basic framework for the mu-calculus of domain logic. The mu-calculus for integers has been shown to be complete. However, it is not clear how that technique could be generalized to higher types, although the theorems about the derivable formulae should be helpful for the proof of completeness of higher types. It is desirable to get completeness and expressiveness results for all closed types. It would also be interesting to study what kind of properties are expressed by the mu-calculi of other interesting types.

5. The work of Chapter 5 can be extended easily to include the powerdomains and modalities. A treatment of Hennessy-Milner logic with recursion via the mu-calculus is called for. Proof systems for satisfiability of such a logic have been given by Larsen ([La90]), but no work on the mu or nu-calculi of Hennessy-Milner logic has been reported yet.

6. To get more expressive power, further extending of the mu-calculus is necessary. It is tempting to add negation and nu-operator to express maximum fixed-points. However, that necessarily leads us outside the Scott open sets, and it is not clear what kind of sets should be used (Smyth proposed the G_δ sets in [Sm83]).

7. An immediate question about the logic of coherent spaces is whether it has anything to do with linear logic since coherent spaces are used as a semantics for linear logic. Notice, however, there is a mismatch here. In linear logic each proposition is interpreted as a coherent space while in the logic of coherent spaces each assertion is interpreted as a stable neighborhood of a certain coherent space. Resolving this mismatch would be a first step towards the understanding of the relationship, if there is any, between linear logic and the domain-logic of coherent spaces.

8. It would be interesting to introduce a language of morphism terms for coherent spaces and stable domains, and to establish the corresponding logical frameworks so as to make it possible to express certain kinds of 'Hoare triples' and 'dynamic logic'. It might be sensible to formulate mu-calculi for the logic of coherent spaces and stable domains.

9. Although some potential difficulties in formulating a logic of partially synchronous morphisms was pointed out, one can still try to get a logic for $\mathbf{SEV}_{syn}$ and $\mathbf{SEV}^*_{syn}$, possibily with a limited form of quantification over assertions. Such a logic might be helpful in understanding the logic of event structures and CCS-like languages.

10. There is some recent progress on the algebraic aspects of stable neighborhoods [EM91]. It might give us some hint in giving a different treatment of the logic of stable domains closer to the alternative suggested in the last section of Chapter 9.

We feel that logics of domains is an important area. It is unlikely that one or two persons can get all the interesting work done in this area. Thus readers are encouraged to work on these and related topics.

Bibliography

[Ab87] Abramsky, S., Domain theory in logical form, Proc. of the 2ed Annual Symposium on Logic in Computer Science. Revised version appeared in Annals of pure and applied logic, 51 (1991).

[Ab87a] Abramsky, S., A domain equation for bisimulation, to appear.

[Ap81] Apt, K.R., Ten years of Hoare 's logic Part I, ACM Trans. on Programming Languages and Systems, 3 (1981).

[Ap84] Apt, K.R., Ten years of Hoare's logic Part II: Nondeterminism , Theoretical Computer Science 28 (1984).

[Ba80] de Bakker, J. W., *Mathematical theory of program correctness,* Prentice Hall International, (1980).

[Ba77] Barwise, K. J., An introduction to first order logic, *The handbook of mathematical logic,* K. J. Barwise, editor, North-Holland, (1977).

[Ba84] Barendregt, H., *The λ-calculus: Its syntax and semantics,* North-Holland, (1984).

[Be78] Berry, G., Stable models of typed λ-calculi, Lecture Notes in Computer Science 62 (1978).

[BCL85] Berry, G., Curien, P.-L. and Levy, J.-J., Full abstraction for sequential languages: the state of art, *Algebraic methods in semantics,* Nivat, M. and Reynolds, J.C. ed, Cambridge University Press (1985).

[Br85] Brookes, S.D., An axiomatic treatment of a parallel programming language, Lecture Notes in Computer Science 193 (1985).

[CGW89] Coquand, T., Gunter, C., Winskel, G., Domain-theoretic models of polymorphism, Information and Computation, 81 (1989).

[Co78] Cook, S., Soundness and completeness of an axiom system for program verification, SIAM J. Computing 7 (1) (1978).

[Cu86] Curien, P.-L., *Categorical combinators, sequential algorithms and functional programming*, Research Notes in Theoretical Computer Science, Pitman, London (1986).

[Di76] Dijkstra, E.W., *A discipline of programming, Prentice-Hall,* (1976).

[Die76] Diers, Y, *Catégories Localisables*, thèse de doctorat d'éetat, Paris VI (1976).

[DG90] Droste, M., and Gobel, R., Universal domains in the theory of denotational semantics of programming languages, Proc. of the IEEE 5th annual symposium on logic in computer science, (1990).

[DG90a] Droste, M., and Gobel, R., Non-deterministic information systems and their domains, Theoretical Computer Science, 75 (1990).

[EM91] Ehrhard, T., Malacaria, P., Stone duality for stable functions, manuscript (1991).

[Gi87a] Girard, J.-Y., The system F of variable types, fifteen years later, Theoretical Computer Science 45 (1986).

[Gi87b] Girard, J.-Y., Linear Logic, Theoretical Computer Science 50 (1987).

[Gi89] Girard, J.-Y., Lafont Y., and Taylor, P., *Proofs and Types*, Cambridge Tracts in Theoretical Computer Science 7 (1989).

[Go79] Gordon, M.J.C., *The denotational description of programming languages*, Springer Verlag (1979).

[Gu85] Gunter, C., *Profinite solutions for recursive domain equations*, PhD thesis, Department of Computer Science, Carnegie-Mellon University, (1985).

[Gu87] Gunter, C., Universal profinite domains, Information and Computation 72, (1987).

[HeMi79] Hennessy, M.C.B., Minler, R., On observing nondeterminism and concurrency, Lecture Notes in Computer Science 85 (1979).

[Ho69] Hoare, C.A.R., An axiomatic basis for computer programming, CACM 12, (1969).

[Ho78] Hoare, C.A.R., Communicating sequential processes, CACM 21, (1978).

[Hy81] Hyland, J.M.E., Function space in the category of locales, Continuous Lattices, Lecture Notes in Mathematics 871 (1981).

[Jo77] Johnstone, P.T., A syntactic approach to Diers' localizable categories, Lecture Notes in Mathematics 753 (1977).

[Jo82] Johnstone, P.T., *Stone spaces*, Cambridge University Press (1982).

[KP78] Kahn, G., Plotkin, G., Domaines concrets, Rapport INRIA Laboria No. 336 (1978).

[Ko83] Kozen, D., Results on the propositional μ-calculus, Theoretical Computer Science 27 (1983).

[LaWi84] Larsen, K.G., Winskel, G., Using information systems to solve recursive domain equations effectively, Lecture Notes in Computer Science 173 (1984).

[La90] Larsen, K.G., Proof systems for satisfiability in Hennessy-Minler logic with recursion, Theoretical Computer Science 72 (1990).

[Ma71] MacLane, S., *Categories for the working mathematician*, Springer-Verlag (1971).

[Me88] Meyer, A. R., Semantical paradigms: Notes for an invited lecture, Third annual symposium on Logic in Computer Science, Edinburgh, Scotland, (1988).

[Mi77] Milner, R., Fully abstract models of typed lambda-calculi, Theoretical Computer Science 4, (1977).

[Mi80] Milner, R., *A calculus of communication systems*, Lecture Notes in Computer Science 92 (1980).

[NiPlWi79] Nielsen, M., Plotkin, G., Winskel, G., Petri nets, event structures and domains, part 1, Theoretical Computer Science 13 (1981).

[NiHe83] de Nicola, R. Hennessy, M.C.B., Testing equivalences for processes Lecture Notes in Computer Science 154 (1983).

[OwGr76] Owicki, S.S., Gries, D., An axiomatic proof technique for parallel programs, Acta Informatica 6 (1976).

[Ph90] Phoa, W., Effective domains and intrinsic structure, Proc. of the IEEE 5th annual symposium on logic in computer science, (1990).

[Pl76] Plotkin, G., A powerdomain construction, SIAM J. Computing 5 (1976).

[Pl78] Plotkin, G., *The category of complete partial orders: a tool for making meanings*, (Pisa Notes) Proc. Summer School on Foundations of Artificial Intelligence and Computer Science, Instituto di Scienze dell' Informazione, Universita di Pisa (1978).

[Pl78a] Plotkin, G., T^{ω} as a universal domain, Journal of Computer and System Sciences 17, (1978).

[Pl80] Plotkin, G., Dijkstra's predicate transformers and Smyth's powerdomains, Lecture Notes in Computer Science 86 (1980).

[Pl81] Plotkin, G.D., A structural approach to operational semantics DAIMI Report FN-19, Aarhus University (1981).

[Pn77] Pnueli, A., The temporal logic of programs, Proc. 19th Annual Symposium on the Foundations of Computer Science (1977).

[Pr79] Pratt, V.R., Dynamic Logic, Proc. 6th International Congress of Logic, Methodology and Philosophy of Science, Hannover (1979).

[Ro86] Robinson, E., Power-domains, Modalities and Vietoris Monad, Technical Report 98, Computer Laboratory, University of Cambridge, (1986).

[Ro87] Robinson, E., Logical aspects of denotational semantics, Lecture Notes in Computer Science 283 (1987).

[Sc86] Schmidt, D., *Denotational semantics*, Allyn and Bacon, Inc., (1986).

[Sc76] Scott, D., Data types as lattices, SIAM Journal of Computing 5 (1976).

[Sc81] Scott, D., Lectures on a mathematical theory of computation, Oxford University Computing Laboratory Technical Monograph PRG-19 (1981).

[Sc82] Scott, D. S., Domains for denotational semantics, Lecture Notes in Computer Science 140 (1982).

[ScSt71] Scott, D.S., Strachey, C., Towards a mathematical semantics of computer languages, Technical Report PRG-6, Oxford University Computing Laboratory, Programming Research Group (1971).

[Sh87] Shawe-Taylor, J. S., *The semantics and open set analysis of a first order programming language*, MSc thesis, Imperial College of Science and Technology (1987).

[SmPl82] Smyth, M.B., Plotkin, G.D., The category-theoretic solution of recursive domain equations, SIAM Journal of Computing, vol. 11 (1982).

[Sm78] Smyth, M.B., Powerdomains, Journal of Computer and System Sciences 16, (1978).

[Sm83] Smyth, M.B., Power domains and predicate transformers: a topological view. Lecture Notes in Computer Science 154 (1983).

[Sm83b] Smyth, M.B., The largest cartesian closed category of domains, Theoretical Computer Science 27, (1983).

[St88] Stoughton, A., *Fully abstract models of programming languages*, Research Notes in Theoretical Computer Science, Pitman, London (1988).

[St77] Stoy, J.E., *Denotational semantics*, MIT press, (1977).

[Ta55] Tarski, A., A lattice-theoretical fixed point theorem and its applications, Pacific Journal of Mathematics, vol. 5 (1955).

[Ta90] Taylor, P., An algebraic approach to stable domains, Journal of pure and applied algebra, 64 (1990).

[Vi88] Vicker, S. *Logic via topology,* Cambridge University Press, (1989).

[Wi80] Winskel, G., *Events in computation,* PhD thesis, University of Edinburgh (1980).

[Wi82] Winskel, G., Event structures semantics of CCS and related languages, Lecture Notes in Computer Science 140 (1982).

[Wi83] Winskel, G., On powerdomains and modality, Lecture Notes in Computer Science 158, (1983).

[Wi84] Winskel, G., A complete proof system for SCCS with modal assertions, Technical Report 78 , Computer Laboratory, University of Cambridge, (1984).

[Wi88] Winskel, G., An introduction to event structures, Lecture Notes in Computer Science 354 (1988).

[Wi90] Winskel, G., *Introduction to the Formal Semantics of Programming Languages,* MIT press, to appear.

[Zh] Zhang, G.Q., *Logics of Domains,* Ph.D. thesis, University of Cambridge (1989).

[Zh89] Zhang, G.Q., DI-domains as information systems, ICALP-1989, Italy. Revised version to appear in Information and Computation (1989).

[Zh90a] Zhang, G.Q., Stable neighbourhoods, to appear in Theoretical Computer Science (1990).

[Zh90b] Zhang, G.Q., A representation of SFP, submitted (1990).

[Zh90c] Zhang, G.Q., When is maximal total? Technical Report 90-002, Department of Computer Science, University of Georgia, submitted with title 'Maximal Elements in Stable Domains' (1990).

[Zh91] Zhang, G.Q., A monoidal closed category of event structures, proceedings of the 7-th conference on Mathematical Foundations of Programming Semantics, Pittsburgh, (1991).

Index

Progress in Theoretical Computer Science

Editor
Ronald V. Book
Department of Mathematics
University of California
Santa Barbara, CA 93106

Editorial Board
Erwin Engeler
Mathematik
ETH Zentrum
CH-8092 Zurich, Switzerland

Gérard Huet
INRIA
Domaine de Voluceau-Rocquencourt
B. P. 105
78150 Le Chesnay Cedex, France

Jean-Pierre Jouannaud
Laboratoire de Recherche
en Informatique Bât. 490
Université de Paris-Sud
Centre d'Orsay
91405 Orsay Cedex, France

Robin Milner
Department of Computer Science
University of Edinburgh
Edinburgh EH9 3JZ, Scotland

Maurice Nivat
Université de Paris VII
2, place Jussieu
75251 Paris Cedex 05
France

Martin Wirsing
Universität Passau
Fakultät für Mathematik
und Informatik
Postfach 2540
D-8390 Passau, Germany

Progress in Theoretical Computer Science is a series that focuses on the theoretical aspects of computer science and on the logical and mathematical foundations of computer science, as well as the applications of computer theory. It addresses itself to research workers and graduate students in computer and information science departments and research laboratories, as well as to departments of mathematics and electrical engineering where an interest in computer theory is found.

The series publishes research monographs, graduate texts, and polished lectures from seminars and lecture series. We encourage preparation of manuscripts in some form of TeX for delivery in camera-ready copy, which leads to rapid publication, or in electronic form for interfacing with laser printers or typesetters.

Proposals should be sent directly to the Editor, any member of the Editorial Board, or to: Birkhäuser Boston, 675 Massachusetts Avenue, Cambridge, MA 02139.

NEW IN 1991:

1. Leo Bachmair, *Canonical Equational Proofs*
2. Howard Karloff, *Linear Programming*
3. Ker-I Ko, *Complexity Theory of Real Functions*
4. Guo-Qiang Zhang, *Logic of Domains*